Prevention through Design

Mechanical–Electrical Systems

Instructor's Manual

DEPARTMENT OF HEALTH AND HUMAN SERVICES
Centers for Disease Control and Prevention
National Institute for Occupational Safety and Health

Disclaimer

Mention of any company or product does not constitute endorsement by NIOSH. In addition, citations to Web sites external to NIOSH do not constitute NIOSH endorsement of the sponsoring organizations or their programs or products. Furthermore, NIOSH is not responsible for the content of these Web sites.

Ordering Information

This document is in the public domain and may be freely copied or reprinted. To receive NIOSH documents or other information about occupational safety and health topics, contact NIOSH at

   Telephone: 1–800–CDC–INFO (1–800–232–4636)
   TTY: 1–888–232–6348
   Web site: www.cdc.gov/info

or visit the NIOSH Web site at www.cdc.gov/niosh

For a monthly update on news at NIOSH, subscribe to *NIOSH eNews* by visiting www.cdc.gov/niosh/eNews.

DHHS (NIOSH) Publication No. 2013–134

August 2013

SAFER • HEALTHIER • PEOPLE™

Please direct questions about these instructional materials to the National Institute for Occupational Safety and Health (NIOSH):

Telephone: (513) 533–8302
E-mail: preventionthroughdesign@cdc.gov

Foreword

A strategic goal of the Prevention through Design (PtD) Plan for the National Initiative is for designers, engineers, machinery and equipment manufacturers, health and safety (H&S) professionals, business leaders, and workers to understand the PtD concept. Further, they are to apply these skills and this knowledge to the design and redesign of new and existing facilities, processes, equipment, tools, and organization of work. In accordance with the PtD Plan, this module has been developed for use by educators to disseminate the PtD concept and practice within the undergraduate engineering curricula.

John Howard, M.D.
Director, National Institute for
    Occupational Safety and Health
Centers for Disease Control and Prevention

Contents

Acknowledgments

Authors:

James McGlothlin, MPH, Ph.D., CPE
John R. Weaver
Anna Menze

The authors thank the following for their reviews:

NIOSH Internal Reviewers

Pamela E. Heckel, Ph.D., P.E.
Donna S. Heidel, M.S., C.I.H.
Thomas J. Lentz, Ph.D., M.P.H.
Rick Niemeier, Ph.D.
Andrea Okun, Ph.D.
Paul Schulte, Ph.D.
Pietra Check, M.P.H.
John A. Decker, Ph.D.
Matt Gillen, M.S., C.I.H.
Roger Rosa, Ph.D.

Peer and Stakeholder Reviewers

Don Bloswick, Ph.D.
COL Daisie D. Boettner, Ph.D.
Joe Fradella, Ph.D.
Matthew Marshall, Ph.D.
Gopal Menon, P.E.
James Platner, Ph.D.
Georgi Popov, Ph.D.
Deborah Young-Corbett, Ph.D., C.I.H., C.S.P., C.H.M.M.

Introduction

This Instructor's Manual is part of a broad-based multi-stakeholder initiative, Prevention through Design (PtD). This module has been developed for use by educators to disseminate the PtD concept and practice within the undergraduate engineering curricula. PtD anticipates and minimizes occupational safety and health hazards and risks* at the design phase of products,† considering workers through the entire life cycle from the construction workers to the users, maintenance staff, and, finally, the demolition team. The engineering profession has long recognized the importance of preventing occupational safety and health problems by designing out hazards. Industry leaders want to reduce costs by preventing negative safety and health consequences of poor designs. Thus, owners, designers, and trade contractors all have an interest in the final design.

This manual is for one of four PtD education modules to increase awareness of construction hazards. The modules support undergraduate courses in civil and construction engineering. The four modules cover the following:

1. Reinforced concrete design
2. **Mechanical–electrical systems**
3. Structural steel design
4. Architectural design and construction.

This manual is specific to a PowerPoint slide deck related to Module 2, **Mechanical–electrical systems**. It contains learning objectives, slide-by-slide lecture notes, case studies, test questions, and a list of citations. It is assumed that the users are experienced professors/lecturers in schools of engineering. As such, the manual does not provide specifics on how the materials should be presented. However, background insights are described for most of the slides for the instructor's consideration.

Numerous examples of inadequate design and catastrophic failure can be found on the Internet. If time permits, have the students seek, share, and analyze appropriate and inadequate designs. The PtD Web site is located at www.cdc.gov/niosh/topics/ptd. The National Institute for Occupational Safety and Health (NIOSH) Fatality Assessment and Control Evaluation (FACE) Reports can be found at www.cdc.gov/niosh/face/. Occupational Safety and Health Administration (OSHA) Fatal Facts are available at www.setonresourcecenter.com/MSDS_Hazcom/FatalFacts/index.htm.

---

*A "hazard" is anything with the potential to do harm. A "risk" is the likelihood of potential harm from that hazard being realized.

†The term *products* under the PtD umbrella pertains to structures, work premises, tools, manufacturing plants, equipment, machinery, substances, work methods, and systems of work.

Learning Objectives and Overview

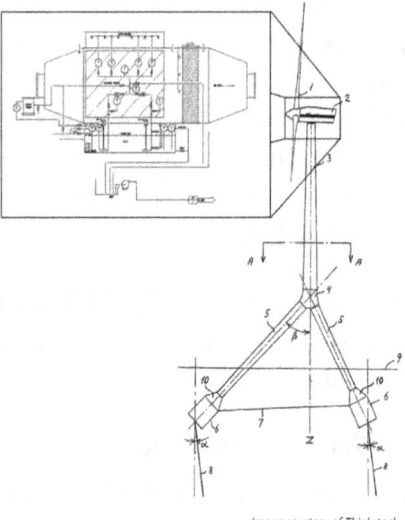

Image courtesy of Thinkstock

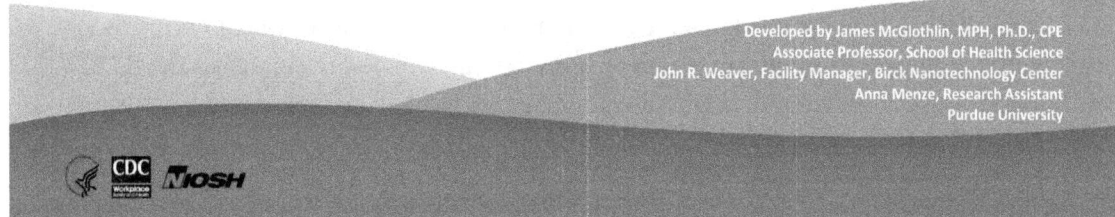

Mechanical–Electrical Systems

EDUCATION MODULE

Developed by James McGlothlin, MPH, Ph.D., CPE
Associate Professor, School of Health Science
John R. Weaver, Facility Manager, Birck Nanotechnology Center
Anna Menze, Research Assistant
Purdue University

NOTES TO INSTRUCTORS

This module contains specific examples of common workplace hazards related to mechanical-electrical systems and illustrates the safety features built into the systems. Actual FACE reports are referenced to aid in leading discussions. One section of slides presents the Prevention through Design (PtD) concept. Applications of the PtD concept to real-world scenarios are presented in case studies.

This education module is intended to facilitate incorporation of the PtD concept into your mechanical-electrical systems design course. You may wish to supplement the information presented in this module and may assign projects, class presentations, or homework as time permits. Sections may be presented independently of the whole. Presentation times are approximate, based on our presentation experience.

To activate the features of some slides, please "enable content," make this a "trusted document," and view the slides in "slide show" mode. To show the presentation file in slideshow mode, press F5. Each slide is accompanied by speaker notes that you can read aloud while the slide is projected on the screen. The audience does not see the speaker notes. When you click on "Use Presenter View" on the Slide Show tab, your monitor displays the speaker notes but the projected image does not.

Thank you for using this module. To report problems or to make suggestions, please contact the National Institute for Occupational Safety and Health (NIOSH):

Telephone: (513) 533–8302
E-mail: preventionthroughdesign@cdc.gov

SOURCE
Image courtesy of Thinkstock

PtD Guide for Instructors

Topic	Slide numbers	Approx. minutes
Introduction to Prevention through Design	5–29	45
Electrical Hazards	30–36	10
Wind Farm Case Study	37–42	10
Nanotechnology Laboratory	43–59	50
Recap	60–61	5
References and Other Sources	62–64	—

Mechanical-Electrical

NOTES

The first two slides of the presentation provide acknowledgments and general information. Learning objectives are delineated on Slide 3. Slide 4 contains the Overview. Slides 5 through 29 discuss construction hazards and introduce the PtD concept and can be covered in approximately 45 minutes. The National Occupational Research Agenda (NORA) has identified three strategic goals for persons working with electricity. They are covered on Slides 30 through 36. Slides 37 through 42 cover the fall protection systems used at the Bowen Wind Farm. Slides 43 through 59 contain pictures and five video clips about PtD concepts embedded into the design of a state-of-the-art center for nanotechnology research. PtD concepts are summarized on slide 60.

 **Learning Objectives**

- Explain the Prevention through Design (PtD) concept.

- List reasons why project owners may wish to incorporate PtD in their projects.

- Identify workplace hazards and risks associated with design decisions and recommend design alternatives to alleviate or lessen those risks.

NOTES

After completing this education module, you should be able to do the following:

- Explain the PtD concept
- Describe motivations, barriers, and enablers for implementing PtD in projects
- List three reasons why PtD improves business value.

 **Overview**

- PtD Concept

- Wind Farm

- Nanotechnology Laboratory

Photo courtesy of Thinkstock

NOTES

This is an overview of the PtD topics that we will cover. Many of you are probably not familiar with PtD, so we will spend a few minutes discussing the concept. Next we will identify safety features of specific mechanical-electrical systems. There are two case studies:

- Fall protection systems at a wind farm
- Five safety features at a laboratory conducting nanotechnology research

SOURCE

Photo courtesy of Thinkstock

Introduction to Prevention through Design (PtD)

NOTES

Let's start by introducing PtD.

Occupational Safety and Health

- Occupational Safety and Health Administration (OSHA)
  www.osha.gov
    - Part of the Department of Labor
    - Assures safe and healthful workplaces
    - Sets and enforces standards
    - Provides training, outreach, education, and assistance
    - State regulations possibly more stringent

- National Institute for Occupational Safety and
  Health (NIOSH) www.cdc.gov/niosh
    - Part of the Department of Health and Human Services, Centers
      for Disease Control and Prevention
    - Conducts research and makes recommendations for the
      prevention of work-related injury and illness

NOTES

All employers, including structural design firms, are required by law to provide their employees with a safe work environment and training to recognize hazards that may be present. They also must provide equipment or other means to minimize or manage the hazards.

Designers historically have not been familiar with the federal Occupational Safety and Health Act (OSH Act) standards because they were rarely exposed to construction jobsite hazards. However, with the increasing roles that designers are playing on worksites, such as being part of a design-build team, it is becoming increasingly important that they receive construction safety training, including information about federal and state construction safety standards.

The Occupational Safety & Health Act of 1970, Public Law 91-596 (OSH Act) [29 USC* 1900], was passed on December 29, 1970, "To assure safe and healthful working conditions for working men and women; by authorizing enforcement of the standards developed under the Act; by assisting and encouraging the States in their efforts to assure safe and healthful working conditions; by providing for research, information, education, and training in the field of occupational safety and health; and for other purposes." The construction industry standards

---

*United States Code. See USC in Sources.

enforced by the Occupational Safety and Health Administration (OSHA) are found in Title 29 Part 1926 of the Code of Federal Regulations [29 CFR 1926].

The National Institute for Occupational Safety and Health (NIOSH) is part of the Department of Health and Human Services, Centers for Disease Control and Prevention. The National Occupational Research Agenda (NORA) is a partnership program to stimulate innovative research and improved workplace practices. Unveiled in 1996, NORA has become a research framework for NIOSH and the nation. Diverse parties collaborate to identify the most critical issues in workplace safety and health. Partners, then, work together to develop goals and objectives for addressing these needs. Participation in NORA is broad, including stakeholders from universities, large and small businesses, professional societies, government agencies, and worker organizations. NIOSH and its partners have formed ten NORA Sector Councils: Agriculture, Forestry & Fishing; Construction; Healthcare & Social Assistance; Manufacturing; Mining; Oil and Gas Extraction; Public Safety; Other Services; Transportation, Warehousing & Utilities; and Wholesale and Retail Trade. The mission of the NIOSH research program for the Construction sector is to eliminate occupational diseases, injuries, and fatalities among individuals working in these industries through a focused program of research and prevention.

SOURCES

CFR. Code of Federal Regulations. Washington, DC: U.S. Government Printing Office, Office of the Federal Register.

NIOSH FACE reports [www.cdc.gov/niosh/face]

OSHA Fatal Facts Accident Reports Index [www.setonresourcecenter.com/MSDS_Hazcom/FatalFacts/index.htm]

OSHA home page [www.osha.gov]

USC. United States Code. Washington, DC: U.S. Government Printing Office.

Construction Hazards

Construction Hazards

- Cuts

- Electrocution

- Falls

- Falling objects

- Heat/cold stress

- Musculoskeletal disease

- Tripping

[BLS 2006; Lipscomb et al. 2006]

Graphic courtesy of OSHA

Mechanical-Electrical

NOTES

A construction worksite by its nature involves numerous potential hazards. A portion of the work is directly affected by weather. Workers interact with heavy equipment and materials at elevated heights, in below-ground excavations, and in multiple awkward positions. The composition of the site workforce changes over the project, and work is done autonomously at times and in coordination at others. The construction worksite is unforgiving to poor planning and operational errors.

For these reasons, pre-job construction-phase planning is used as a best practice to systematically address potential hazards. Project-specific worker safety orientations prior to site work also play an important role. PtD practices, by systematically looking further upstream at design-related potential hazards, extend these pre-job measures. PtD can help identify potential hazards so that they can be eliminated, reduced, or communicated to contractors for pre-job planning.

Every hazard that can be addressed should be addressed. Falling can cause serious injury. Boilermakers, pipe-fitters, and iron workers can experience career-ending musculoskeletal injuries by lifting heavy loads or working in a cramped position. Anyone can be seriously injured by a falling object. Whether a structural member or a simple wrench, a falling object can be deadly. Anyone can trip, but the elevated height and proximity to dangerous equipment increase the risk of injury on a construction site. Some accidents are caused by poor lighting and/or sunlight glare. Common injuries due to spatial misperception include hitting your head or cutting yourself on sharp corners. Hot summer and cold winter days can affect worker health. Personal protective equipment (PPE), such as hardhats, gloves, ear protection, and safety glasses, is required for a reason! Not every hazard on a construction worksite can be "designed out," but many significant ones can be minimized during the design phase.

SOURCES

BLS [2006]. Injuries, illnesses, and fatalities in construction, 2004. By Meyer SW, Pegula SM. Washington, DC: U.S. Department of Labor, Bureau of Labor Statistics, Office of Safety, Health, and Working Conditions [www.bls.gov/opub/cwc/sh20060519ar01p1.htm].

Lipscomb HJ, Glazner JE, Bondy J, Guarini K, Lezotte D [2006]. Injuries from slips and trips in construction. Appl Ergonomics 37(3):267–274.

OSHA [ND]. Fatal Facts Accident Reports Index [foreman electrocuted]. Accident summary no. 17 [www.setonresourcecenter.com/MSDS_Hazcom/FatalFacts/index.htm].

Graphic Courtesy of OSHA

ACCIDENT SUMMARY No. 17

Accident Type:	Electrocution	
Weather Conditions:	Sunny, Clear	
Type of Operation:	Steel Erection	
Size of Work Crew:	3	
Collective Bargaining	No	
Competent Safety Monitor on Site:	Yes - Victim	
Safety and Health Program in Effect:	No	
Was the Worksite Inspected Regularly:	Yes	
Training and Education Provided:	No	
Employee Job Title:	Steel Erector Foreman	
Age & Sex:	43-Male	
Experience at this Type of Work:	4 months	
Time on Project:	4 Hours	

BRIEF DESCRIPTION OF ACCIDENT

Employees were moving a steel canopy structure using a "boom crane" truck. The boom cable made contact with a 7200 volt electrical power distribution line electrocuting the operator of the crane; he was the foreman at the site.

INSPECTION RESULTS

As a result of its investigation. OSHA issued citations for four serious violations of its construction standards dealing with training, protective equipment, and working too close to power lines.

OSHA's construction safety standards include several requirements which, If they had been followed here, might have prevented this fatality.

ACCIDENT PREVENTION RECOMMENDATIONS

1. Develop and maintain a safety and health program to provide guidance for safe operations (29 CFR 1926.20(b)(1)).
2. Instruct each employee on how to recognize and avoid unsafe conditions which apply to the work and work areas (29 CFR 1926.21(b)(2))
3. If high voltage lines are not de-energized, visibly grounded, or protected by insulating barriers, equipment operators must maintain a minimum distance of 10 feet between their equipment and the electrical distribution or transmission lines (29 CFR 1926.550(a)(15)(i)).

SOURCES OF HELP

- Ground Fault Protection on Construction Sites (OSHA 3007) which describes OSHA requirements for electrical safety at construction sites.

- Construction Safety and Health Standards (OSHA 2207) which contains all OSHA job safety and health rules and regulations (1926 and 1910) covering construction
- OSHA Safety and Health Training Guidelines for Construction (available from the National Technical Information Service - Order No PB-239312/AS) comprised of a set of 15 guidelines to help construction employees establish a training program in the safe use of equipment, tools, and machinery on the job
- OSHA-funded free onsite consultation services Consult your telephone directory for the number of your local OSHA area or regional office for further assistance and advice (listed under the US Labor Department or under the state government section where states administer their own OSH programs).

NOTE: The case here described was selected as being representative of fatalities caused by improper work practices. No special emphasis or priority is implied nor is the case necessarily a recent occurrence. The legal aspects of the incident have been resolved, and the case is now closed.

Construction Accidents

 **Construction Accidents in the United States**

Construction is one of the most hazardous occupations. This industry accounts for

- 8% of the U.S. workforce, but 20% of fatalities

- About 1,100 deaths annually

- About 170,000 serious injuries annually

[CPWR 2008]

Photo courtesy of Thinkstock

Mechanical-Electrical

NOTES

As many of us know, construction is one of the most dangerous industries for workers. In the United States, construction typically accounts for 170,000 serious injuries and 1,100 deaths each year. The fatality rate is disproportionally high for the size of the construction workforce. Twenty percent of all collapses during construction are the result of structural design errors. Statistics like these do not tell the whole story. Behind every serious injury, there is a real story of an individual who suffered serious pain and may never fully recover. Behind every fatality, there are spouses, children, and parents who grieve every day for their loss. We all recognize that safety is a vital component of an inherently dangerous business. All of us—including architects and engineers—must do what we can to reduce the risk of injuries on our projects.

SOURCES

CPWR [2008]. The construction chart book. 4th ed. Silver Spring, MD: Center for Construction Research and Training.

New York State Department of Health [2007]. A plumber dies after the collapse of a trench wall. Case report 07NY033 [www.cdc.gov/niosh/face/pdfs/07NY033.pdf].

OSHA [ND]. Fatal Facts Accident Reports Index [laborer struck by falling wall]. Accident summary no. 59 [www.setonresourcecenter.com/MSDS_Hazcom/FatalFacts/index.htm].

Photo courtesy of Thinkstock

NEW YORK

state department of

HEALTH

FATALITY ASSESSMENT AND CONTROL EVALUATION

A Plumber Dies After the Collapse of a Trench Wall
Case Report: 07NY033

SUMMARY

In May 2007, a 46 year old self-employed plumbing contractor (the victim) died when the unprotected trench he was working in collapsed. The victim was an independent plumber subcontracted to install a sewer line connection to the sewer main, part of a general contractor project to install a new sanitary sewer for an existing single family residence.

At approximately 12:30 PM on the day of the incident, the workers on site observed the victim walking back toward the residence for parts as they initiated their lunch break. When the victim did not come for his lunch or answer his cell phone, the general contractor and workers starting searching for the victim. The excavation contractor observed that a portion of the trench had collapsed where the victim was installing a sewer tap. The victim was found trapped in the trench under a large slab of asphalt, rock and soil. Three workers immediately climbed down the side of the trench to try to assist the victim. One of the workers called 911 on his cell phone. Police and emergency medical services (EMS) arrived on site within minutes. The EMS members entered the unprotected trench but could not revive the victim. The county trench rescue team recovered the victim's body at approximately seven feet below grade and lifted him from the ditch four hours after the incident. He was pronounced dead at the site. More than 50 rescue workers were involved in the recovery.

New York State Fatality Assessment and Control Evaluation (NY FACE) investigators concluded that, to help prevent similar occurrences, employers and independent contractors should:

- **Require that all employees, subcontractors, and site workers working in trenches five feet or more in depth are protected from cave-ins by an adequate protection system.**
- **Require that a competent person conducts daily inspections of the excavations, adjacent areas, and protective systems and takes appropriate measures necessary to protect workers.**
- **Require that all employees and subcontractors have been properly trained in the recognition of the hazards associated with excavation and trenching. In addition, the general contractor (GC) should be responsible for the collection and review of training records and require that all workers employed on the site have received the requisite training to meet all applicable standards and regulations for the scope of work being performed.**
- **Require that on a multi-employer work site, the GC should be responsible for the coordination of all high hazard work activities such as excavation and trenching.**

- **Require that all employees are protected from exposure to electrical hazards in a trench.**

Additionally,

- **Employers of law enforcement and EMS personnel should develop trench rescue procedures and should require that their employees are trained to understand that they are not to enter an unprotected trench during an emergency rescue operation.**
- **Local governing bodies and codes enforcement officers should receive additional training to upgrade their knowledge and awareness of high hazard work, including excavation and trenching. This skills upgrade should be provided to both new and existing codes enforcement officers.**
- **Local governing bodies and codes enforcement officers should consider requiring building permit applicants to certify that they will follow written excavation and trenching plans in accordance with applicable standards and regulations, for any projects involving excavation and trenching work, before the building permits can be approved.**

INTRODUCTION

In May, 2007, a 46 year old self-employed plumbing contractor died when the trench he was working in collapsed at a residential construction site. Approximately 8000 pounds of broken asphalt, rock and dirt fell atop the victim, fatally crushing him as he was installing a sewer tap to a town sewer main. The New York State Fatality Assessment and Control Evaluation (NY FACE) program learned about the incident from a newspaper article the following day. The Occupational Safety and Health Administration (OSHA) investigated the incident along with the county sheriff's office. The NY FACE staff met and reviewed the case information with the OSHA compliance officer. This report was developed based upon the information provided by OSHA, the county sheriff's department, and the county coroner's medical and toxicological reports.

The general contractor (GC) on the residential construction site had been hired by the homeowners to complete a project that included the installation of a new sanitary sewer connection for an existing single family residence. The GC was the owner and sole employee of his company, which had been in business for many years. The GC directed the work of two subcontractors on the work site to complete the installation of the residential sewer line.

- One subcontractor was an excavating company that had been in business for approximately four years. The owner of this company hired two workers to assist him with the excavation of the trench.
- The second subcontractor was the victim, a self-employed licensed plumber who had over twenty years of experience with a variety of construction projects, including the installation of sewer lines. The victim did not have any previous work relationship with either the GC or the excavation subcontractor.

The OSHA investigation report indicated that the GC and the subcontractor did not have health and safety programs. A formal health and safety plan had not been established to identify the hazards of the excavation project and the actions to be taken to remediate them. The GC, subcontractors and the subcontractors' employees did not have hazard recognition training or safety training on the fundamentals of excavation and trenching. None of the workers on the site were knowledgeable on excavation and trenching safety standards and applicable regulations and they did not understand the hazards and dangers associated with working in a trench. A competent person was not present to conduct initial and ongoing inspections of the excavation project, identify potential health and safety hazards such as possible cave-in, and oversee the use of adequate protection systems and work practices.

INVESTIGATION

The GC was hired to replace a crushed sewer line that ran under the driveway of an existing single family residence. Rather than dig up the driveway to replace the old line, which was thought to be more costly and time-consuming, the GC decided to run a new line. All required town permits had been obtained and the local codes enforcement requirements for one-call system notification of the excavation and underground utility location mark-outs had been completed. The work had been scheduled to be completed in one day (Friday), but the excavation subcontractor lost time due to hitting a water line and encountering very rocky soil during the excavation. The project had to be extended to two days (Friday and Monday). The town water and sewer inspector visited the work site on Friday, observed the digging of the trench which began at the residence, and halted the digging of the trench at the edge of the property to avoid having an open trench in the road and consequent road closure over a weekend. Excavation company workers had been observed in the trench spotting and hand digging.

On Monday, the day of the incident, the excavating subcontractor initiated excavation from the edge of the road to the sewer main in the roadway. An employee witness of the excavating company stated that the victim was directing excavation work while in the trench and hand digging to expose the sewer main once the excavator came close to the location. OSHA findings indicated that tools were uncovered in the trench in the area of the trench wall collapse, including a shovel, pick ax, hammer drill and drill bits, consistent with the scenario of the victim being in the ditch, hand digging to locate the sewer main. The town water and sewer inspector also visited the work site on Monday. He determined that the victim did not have the correct parts to complete the sewer connection, advised him of the correct parts, and indicated that he would return later in the day to re-inspect and photograph the completed sewer tap in order to allow the excavating subcontractor to run the pipe back to the house, backfill the excavation and reopen the road.

The GC left the work site to purchase the correct parts, while the excavation continued. The dimensions of the final trench were approximately 55 feet in length, 3 feet to 8 feet in depth, and 30 inches to 128 inches in width (see Figure 1). It was shaped like a "T." The gravity sewer main that the victim was connecting to was located at a depth of 7 feet 4 inches (7' 4") below grade at the east (E) end of the top of the "T." Installation of new sewer pipe from the residence had been initiated and some of the trench had already been backfilled. The length of the trench from the top of the "T" to the location of the newly installed sewer pipe was 35 feet 11 inches (35'11") at the time of the incident. Soil analysis results, conducted after the incident, indicated a granular, sandy gravel Type C soil (OSHA Excavation Standard) that contained large cobbles and boulders, the least stable soil type.

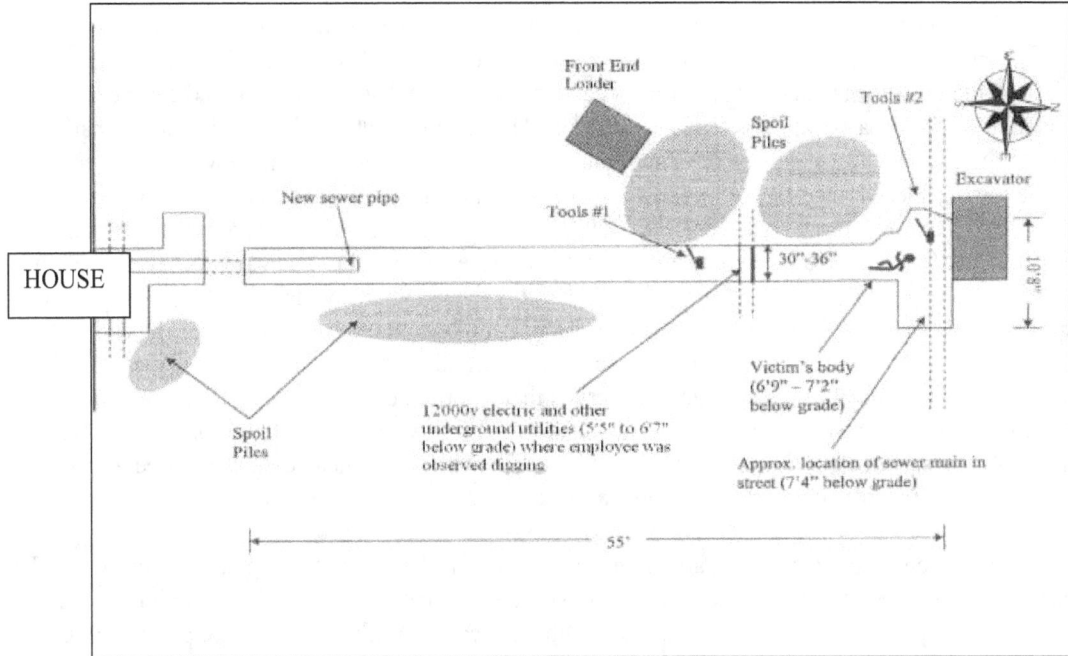

Figure 1: Schematic of the excavation and the incident site (courtesy of OSHA)

The faces of the trench were vertical. No shoring or benching was used. Large cobbles and boulders and loose rock/dirt were visible on the face of the excavation and were not removed or supported. The pavement above the E and W faces of the excavation had been undermined during excavation activities and no support system was utilized to protect employees from a possible collapse. Pieces of road pavement and asphalt had been undermined during excavation activities in the road in the proximity of the sewer main at the top of the "T." These areas were in plain view and did not have additional support. On the W side of the excavation, loose boulders, rock and debris in spoils piles were located less than two feet from the edge of the trench. (Figure 2) The excavator was positioned adjacent to the N end of the trench, where undermined areas were in plain sight. The N end of the trench, where the victim was installing the sewer tap, also lacked an access ladder or other safe means of entry/egress.

Figure 2: View of the west wall of the excavation south of the "T."
Note the boulders and loose rock/dirt on the excavation face as well as the location of the spoils pile within 2 feet of the edge of the trench. (courtesy of OSHA)

The GC returned just before 12 noon with the correct parts and handed them to the victim. The GC left the site in order to purchase lunch for the workers, including the victim. At this same time, the victim called the town water and sewer inspector, informed him that he had located the sewer main, had all the correct parts, and was ready to connect. The town inspector informed the victim that someone from the town would be out after lunch to inspect and photograph the sewer tap. According to the town inspector, a sewer tap to a sewer main is a simple job that would take about 20 minutes to complete. The GC returned with lunch at 12:30PM. The workers, with the exception of the victim, took a break for lunch at a location near the front end loader (Figure 1). The workers saw the victim walking in the trench in the direction of the residence and heard him say that he was "looking for a splitter for a three-way." By 1:00 PM the victim still had not come for his lunch. The GC called the victim on his cell phone and looked for him in his van behind the house. The other workers joined in the search. The excavating subcontractor observed that a portion of the west side of the trench had collapsed. When the workers approached the excavation, they found the victim trapped in the trench under a large slab of asphalt, rock, and soil, with only the back of his head exposed. Three workers climbed down the side of the trench to try to assist the victim.

The workers removed the dirt from around his head, lifted his head, and tried to clear his airway. They checked for a pulse, but found none. One of the workers then called 911 from his cell phone. The workers attempted to move the slab of asphalt without success. Within minutes, the police arrived, followed by EMS at approximately 1:08 PM. The EMS personnel entered the unprotected trench but were unable to revive the victim. Volunteer firefighters from multiple fire departments and a special trench rescue team responded, the latter team having been created by the county after the deaths of two workers in a construction trench collapse 10 years earlier. A wooden safety box was built by the trench rescue team and efforts began to free the victim from entrapment by chipping the asphalt slab into pieces. Using a system of ropes and pulleys, the rescue team lifted the victim from the ditch at 4:25 PM. His body had been recovered at about 7' below grade. The county coroner pronounced him dead at 4:35 PM. Approximately 50 rescuers responded to the 911 call.

The OSHA investigation resulted in findings that the trench section that collapsed was a triangular shaped area at the northwest corner of the excavation, approximately 5 feet 1 inch (5' 1") in length, 4 feet (4') wide, and 6-7 feet (6-7') deep. Multiple hazards were present, but had not been identified and remediated. The W side of the excavation collapsed and pieces of asphalt paving and rock fatally crushed the victim while he was making the sewer tap (Figures 3 and 4).

The hazards of the unprotected trench exposed additional people to the excavation collapse as the GC, the excavation company workers and EMS personnel entered the trench to attempt a rescue of the victim. In addition to the trench hazards, no precautions had been taken to prevent exposure to the underground electrical and utility lines. The town inspector had noted that a young employee of the excavation company was "manually hand digging with shovel and pick ax "within a few inches of the buried electrical lines." This is consistent with OSHA findings that indicated attempts had been made to cut the rock in the face of the trench at the location of the underground utilities. A demo saw, hammer drill and cordless reciprocating saw used to cut rocks and pavement were found within inches of the 12,000 volt underground electrical line. Several other utilities were also exposed in this location at the edge of the road (Figure #1, Tools #1). EMS personnel also entered the trench when power was still connected to the utilities in the trench.

Figure 3: Location of collapse.
Note spoils piles and equipment located less
2 feet from the edge of the trench
(courtesy of OSHA)

Figure 4: Area of trench collapse
Note the large boulders hanging from the than
excavation faces and undermined areas on the
edge of the trench (courtesy of OSHA)

RECOMMENDATIONS/DISCUSSION

Recommendation #1: *Employers and independent contractors should require that all employees, subcontractors and site workers working in trenches five feet or more in depth are protected from cave-ins by an adequate protection system.*

Discussion: Employers and contractors should require that all employees working in trenches five feet deep or more are protected from cave-ins by an adequate protection system appropriate to the conditions of the trench, including sloping techniques or support systems such as shoring or trench boxes (OSHA 29CFR 1926.652). Sloping involves positioning the soil away from an excavation trench at an angle that would prevent the soil from caving into the trench. Even in shallow trenches less than five feet in depth, the possibility of accidents still exists. Trenches five feet deep or less should also be protected if a competent person identifies a cave-in potential. Trench protection systems are available to all employers and independent contractors, even as rental equipment. Employers should also require that all pieces of excavated pavement, asphalt, dirt, rock, boulders, and debris as well as excavation equipment are located in spoils piles or positions that are at least two feet from the edge of the excavated trench. Where a two foot setback is not possible, spoils may need to be hauled to another location. In this incident, sloping would not have been an appropriate protection system, due to the composition of the soil. Employers and contractors should consult tables located in the appendices of the OSHA Excavation Standard that detail the protection required based upon the soil type and environmental conditions present at a work site. Employers and contractors can also consult with manufacturers of protective systems to obtain detailed guidance for the appropriate use of protection systems.

Trenches should be kept open only for the minimum amount of time needed. Hinze and Bren (1997) observed that the risk of a collapse in an unprotected trench increases the longer a trench is open. They propose that after a trench is dug, the apparent cohesion of trench walls may begin to relax after only four hours, contributing to increasingly unstable walls in an unprotected trench. In this incident, a 45 feet length of the trench had been excavated and was left open for more than two days. The trench section where the incident occurred was dug at approximately 8:30 AM on the day of the incident. Hand digging and incorrect parts resulted in additional delays in making the sewer tap to the main. The trench collapse occurred approximately four hours later, between 12:30 PM and 1:00 PM.

The key to preventing a trench accident is not to enter an unprotected trench. When the walls of a trench collapse or cave in, the results are entrapment or struck-by incidents to anyone caught inside, accidents which can occur in seconds. Many workers in a trench are in a kneeling or squatting position that results in little opportunity for an escape. Victims do not need to be completely covered in soil. Even with partial covering, enough pressure is created for mechanical asphyxia in which the weight of the dirt and soil compresses the chest. One cubic yard of soil has an average weight of 2500 pounds (Figure 4), but can vary due to the composition and moisture content.

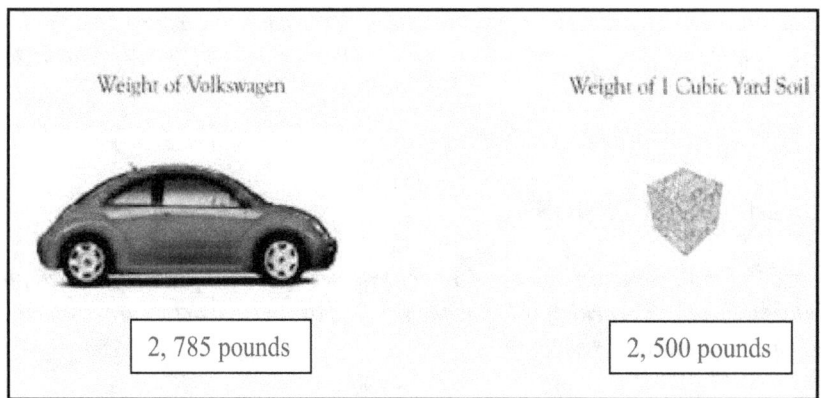

Figure 5: Weight of one cubic yard of soil (courtesy of "Weights of Building Materials, Agricultural Commodities, and Floor Loads for Buildings" standard reference)

Recommendation #2: *Employers and independent contractors should require that a competent person conducts daily inspections of the excavations, adjacent areas, and protective systems and takes appropriate measures necessary to protect workers.*

Discussion: Employers and independent contractors are responsible for complying with the OSHA Excavation Standard requirements to designate a competent person on site for excavation and trenching projects to make daily inspections of excavations, the adjacent areas, and protective systems (OSHA 29CFR 1926.651). A competent person is defined as someone who is capable of identifying existing and predictable hazards in the surroundings and working conditions that are dangerous to employees and who has the authorization to take prompt corrective measures to eliminate them. They should inspect the trenches daily, as needed throughout the work shift, and as conditions change (for example, heavy rainfall or increased traffic vibrations). These inspections should be conducted before worker entry, to ensure that there is no evidence of a possible cave-in, failure of a protective system, hazardous conditions such as spoils piles or equipment location, or hazardous atmosphere.

In particular, competent persons are required by OSHA to complete a competent person training curriculum, which could be an OSHA training program or an equivalent safety or trade organization training. The competent person needs be knowledgeable on the hazards associated with excavation and trenching, as well as the causes of injuries and the safe work practices and specific protective actions needed. Competent persons must also be experienced in excavation and trenching with a minimum of hands-on training in a demonstration trench or in a field component. The competent person needs to know the key points of the OSHA Excavation Standard, including the excavation standards and appendices, checklists, soils analysis and the components of a daily trenching inspection.

Having a competent person is a particularly acute problem among contracting companies that employ fewer than 10 workers. Of the National Institute for Occupational Safety and Health (NIOSH) FACE cases related to excavation and trenching, 88% were non-union companies with less than 10 workers. These small companies are not members of trade associations and are the least likely to employ trench safety protections and to have an adequately trained competent person or an excavation crew.

In this incident, no competent person was hired by the GC to conduct initial and ongoing inspections of the trench. The GC, excavating contractor, and excavation company employees did not possess an understanding of the hazards associated with excavation and trenching operations or a knowledge of the requirements of the OSHA Excavation Standard. No one on-site was qualified to function as the competent person.

Recommendation #3: *Employers and independent contractors should require that all employees and subcontractors have been properly trained in the recognition of the hazards associated with excavation and trenching. On a multi-employer work site, the GC should be responsible for the collection and review of training records and require that all workers employed on the site have received the requisite training to meet all applicable standards and regulations for the scope of work being performed.*

Discussion: Excavation and trenching is one of the most hazardous construction operations. Even with a competent person on site, workers in excavation and trenching operations are also in need of health and safety training, including basic hazard recognition and prevention. Workers should be able to identify the specific hazards associated with excavation and trenching, the reasons for using protective equipment and how to work in a trench safely. Workers should be trained not to enter an unprotected trench, even in a rescue attempt, since they place themselves at risk of becoming injured or killed. If necessary, projects should be delayed until training requirements are met and training records are provided.

In this case, the general contractor, excavation subcontractor, and excavation company employees did not demonstrate adequate knowledge of safe work practices in excavation and trenching. The limited training in proper excavation technique as well as inadequate hazard recognition and prevention training were critical to the failure to properly assess the hazards present and protect the trench.

Recommendation #4: *Employers and independent contractors should require that on a multi-employer work site, the GC should be responsible for the coordination of all high hazard work activities such as excavation and trenching.*

Discussion: The GC is responsible and accountable for the safety of all employees, subcontractors and workers on the site. Health and safety plans should be in place to formally address the hazards that may be encountered, including written plans to manage these hazards and protect the safety of all workers on the site.

In this incident, the GC did coordinate the work activities of the subcontractors and workers on the job, but health and safety plans were not addressed. The management of excavation and trenching hazards was left to a subcontractor who was not a competent person, knowledgeable or trained in the requirements of the OSHA Excavation Standard.

Recommendation #5: *Employers of law enforcement and EMS personnel should develop trench rescue procedures and should require that their employees are trained to understand that they are not to enter an unprotected trench during an emergency rescue operation.*

Discussion: Employers of law enforcement and EMS personnel should develop a formal safety procedure for emergency rescue in an unprotected trench. Entering an unprotected trench after a cave-in or collapse could place would-be rescuers in danger. Rescue is a delicate and slow operation requiring knowledge of the behavior of unstable soil, necessary to prevent further injury to the victim or the rescuers. The added weight and vibrations can also contribute to an increased susceptibility to further collapse. Many rescuers precipitate second and third stage trench cave-ins and have become victims themselves. In this incident EMS personnel entered the unprotected trench in an attempt to rescue the victim, exposing themselves to an excavation collapse hazard.

Emergency rescue workers, such as law enforcement officials and EMS personnel, should receive specialized training in how to rescue workers who may be trapped in utility trenches, and should not put themselves in danger by entering an unprotected trench. In this incident, a specialized rescue team was called in to respond to the emergency. The rescue workers had special equipment for trench rescues and building collapses and had undergone specialized training in the area of trench/building collapse emergencies. They immediately constructed a wooden safety box in the trench with a system of ropes and pulleys before entering the trench to free the victim. National Fire Protection Association (NFPA) 1670, Chapter 11 details the requirements for rescue operations after a trench cave-in occurs.

Recommendation #6: *Local governing bodies and codes enforcement officers should receive additional training to upgrade their knowledge and awareness of high hazard work, including excavation and trenching. This skills upgrade should be provided to both new and existing codes enforcement officers.*

Discussion: This recommendation may create a mechanism of observation and oversight by the codes enforcement officers who are likely to encounter small employers and independent contractors during their work. The officers could inform the employers and contractors of potential hazards, provide fact sheets that highlight the key requirements for the excavation and trenching standards, and check some of the basics of the trenching project such as depth of the trench, protection of the trench and identification of the competent person. In addition, they could advise employers and contractors to contact safety experts to learn about and implement trench safety. This may be an effective accident prevention strategy, reaching the thousands of untrained and unprepared small employers and independent contractors with awareness and guidance, the very workers who represent the major group of fatalities in New York State.

In this incident, the town water and sewer inspector observed workers in the unprotected trench serving as spotters, observed a worker hand digging within a few feet of a live buried electrical utility, and observed the victim spotting in the unprotected trench for the excavating subcontractor while attempting to locate the sewer main. If the above recommendation was in place, with a trained and knowledgeable officer, at a minimum the excavation work may have been halted and entry into an unprotected trench may have been prohibited.

Recommendation #7: *Local governing bodies and codes enforcement officers should consider requiring building permit applicants to certify that they will follow written excavation and trenching plans in accordance with applicable standards and regulations, for any projects involving excavation and trenching work, before the building permits can be approved.*

Discussion: Local governing bodies may consider revising building permits to require building permit applicants to certify that they will follow written plans for any projects involving excavation and trenching. Statements on the permit applications would be added to indicate that the employer/independent contractor agrees to accept and abide by all standards and regulations governing the excavation and trenching work, not just local governing body codes and ordinances. If construction companies and independent contractors were required to provide written documentation of how the high hazard work of excavation and trenching will be performed safely as part of the building permit application process, it may prompt the employers and contractors to plan ahead, formally assess the hazards, seek assistance in developing the required safety and injury prevention program, and implement the necessary injury prevention measures. No work should be initiated unless these requirements are met after review and approval. These changes may help to prevent trench related fatalities in NYS.

Recommendation #8: *Employers and independent contractors should require that all employees are protected from exposure to electrical hazards in a trench.*

Discussion: Utilities to the single family residence were located underground in the trench near the edge of the road. Workers were observed using power and hand tools within inches of live 12,000 volt lines. This did not contribute to the fatality, but did present another potential hazard to workers in the excavation and trenching project and to the rescue workers. Performing cutting work next to hot utility lines could have resulted in additional serious injuries and death from electrocution. The company performed the utility mark-out as required by local codes but did not contact the utility company to turn off the power as required, when they realized the need to hand cut large rocks and boulders in the trench. The power was not shut off to these lines until after the incident, when workers returned to complete the work.

Key words: Trench, collapse, cave-in, trenching, excavation, trench protection systems, entrapment, spoils piles

REFERENCES:

1. Associated General Contractors of America Safety Training for the Focus Four. *Hazards in Construction*. Retrieved February 8, 2011 from http://www.agc.org/cs/career_development/safety_training/focus_four_locations

2. CDC/NIOSH. *NIOSH Safety and Health Topic: Trenching and Excavation*. Retrieved on February 8, 2011 from http://www.cdc.gov/niosh/topics/trenching/

3. CDC/NIOSH. MMWR. 2004. *Occupational Fatalities During Trenching and Excavation Work - United States, 1992-2001*. Morbidity and Mortality Weekly Report, 53(15):311-314. Retrieved February 8, 2011 from www.cdc.gov/mmwr/preview/mmwrhtml/mm5315a2.htm

4. CDC/NIOSH. Alert: July 1985. *Preventing Deaths and Injuries from Excavation Cave-ins*. retrieved February 8, 2011 from http://www.cdc.gov/niosh/85-110.html

5. CDC/NIOSH. Fatality Assessment and Control Evaluation (FACE) investigation reports. Retrieved February 8, 2011 from www.cdc.gov/niosh/face

6. Center to Protect Workers' Rights (CPWR). Plog, Barbara et al. March, 2006. *Barriers to Trench Safety: Strategies to Prevent Trenching-Related Injuries and Deaths*. Retrieved February 8, 2011 from www.elcosh.org.

7. Commonwealth of Massachusetts. Executive Office of Labor and Workforce Development. *Trenching Hazard Alert for Public Works Employers and Employees in Massachusetts*. Bulletin 407, 11/2007, p1-4.

8. Deatherage, J.H., et al. 2004 *Neglecting Safety Precautions may lead to trenching fatalities*. American Journal of Industrial Medicine, 45(6):522-7.

9. EC&M online. June, 2009. *Danger Uncovered*. Beck, Ireland. Retrieved February 8, 2011 from http://ecmweb.com/construction/electrical-trench-safety-20090601/

10. Encyclopedia of Occupational Health and Safety. 4th Edition. *Chapter 93: Construction Trenching* by Jack Mickle. *Types of Projects and Their Associated Hazards* by Jeffrey Hinkman. Retrieved February 8, 2011 from http://www.elcosh.org/en/document/296/d000279/encyclopedia-of-occupational-safety-%2526-health-%253A-chapter-93-construction.html

11. Executive Safety Update. The Monthly News Bulletin of the Construction Safety Center, Vol. 17, Issue 3, September, 2009

12. Hinze, J.W. and K. Bren. 1997. *The causes of trenching-related fatalities*. Construction Congress V: Managing Engineered Construction in Expanding Global Markets. Proceedings of the Congress, sponsored by the American Society of Civil Engineers (ASCE), 131(4): 494-500.

13. Irizarry, J. et al: 2002 *Analysis of Safety Issues in Trenching Operation*. 10th Annual Symposium on Construction Innovation and Global Competitiveness, September 9-13, 2002. Retrieved February 8, 2011 from Construction Safety Alliance site: http://engineering.purdue.edu/CSA/publications/trenching03

14. Job Health and Safety Quarterly. Fall, 2009. *Trenching is a Dangerous and Dirty Business*. Retrieved February 8, 2011 from http://www.elcosh.org/en/document/161/d000168/trenching-is-a-dangerous-and-dirty-business.html

15. Miami-Dade County. *Trench Safety Act Compliance Statement, FM5238 Rev. (12-00)*. Retrieved February 8, 2011 from http://facilities.dadeschools.net/form_pdfs/5238.pdf

16. New York City Department of Buildings. *Excavation and Trench Safety Guidelines* by Dan Eschenasy. www.NYC.gov/buildings. Retrieved February 8, 2011 from http://www.elcosh.org/en/document/161/d000168/trenching-is-a-dangerous-and-dirty-business.html

17. OSHA. *Working Safely in Trenches Safety Tips.* Retrieved February 8, 2011 from http://www.osha.gov/Publications/trench/trench_safety_tips_card.html

18. OSHA. *29CFR1926.650 subpart p. Excavations: scope, application and definitions.* Retrieved February 8, 2011 from http://www.osha.gov/pls/oshaweb/owadisp.show_document?p_id=10774&p_table=STANDARDS

19. OSHA. *29CFR1926.651 subpart p. Excavations: specific excavation requirements.* Retrieved February 8, 2011 from http://www.osha.gov/pls/oshaweb/owadisp.show_document?p_table=STANDARDS&p_id=10775

20. OSHA. *29CFR1926.652 subpart p. Excavations: requirements for protective systems.* Retrieved February 8, 2011 from http://www.osha.gov/pls/oshaweb/owadisp.show_document?p_table=STANDARDS&p_id=10776

21. OSHA. *OSHA Technical Manual SECTION V: CHAPTER 2 EXCAVATIONS: HAZARD RECOGNITON IN TRENCHING AND SHORING.* Retrieved February 8, 2011 from http://www.osha.gov/dts/osta/otm/otm_v/otm_v_2.html

22. OSHA. *OSHA's Construction e-tool.* Retrieved February 8, 2011 from http://www.osha.gov/SLTC/etools/construction/trenching/mainpage.html

The New York State Fatality Assessment and Control Evaluation (NY FACE) program is one of many workplace health and safety programs administered by the New York State Department of Health (NYSDOH). It is a research program designed to identify and study fatal occupational injuries. Under a cooperative agreement with the National Institute for Occupational Safety and Health (NIOSH), the NY FACE program collects information on occupational fatalities in New York State (excluding New York City) and targets specific types of fatalities for evaluation. NY FACE investigators evaluate information from multiple sources and summarize findings in narrative reports that include recommendations for preventing similar events in the future. These recommendations are distributed to employers, workers, and other organizations interested in promoting workplace safety. The NY FACE does not determine fault or legal liability associated with a fatal incident. Names of employers, victims and/or witnesses are not included in written investigative reports or other databases to protect the confidentiality of those who voluntarily participate in the program.

Additional information regarding the NY FACE program can be obtained from:
New York State Department of Health FACE Program
Bureau of Occupational Health
Flanigan Square, Room 230

1-518-402-7900
www.nyhealth.gov/nysdoh/face/face.htm

ACCIDENT SUMMARY No. 59

Accident Type:	Struck by Falling Wall	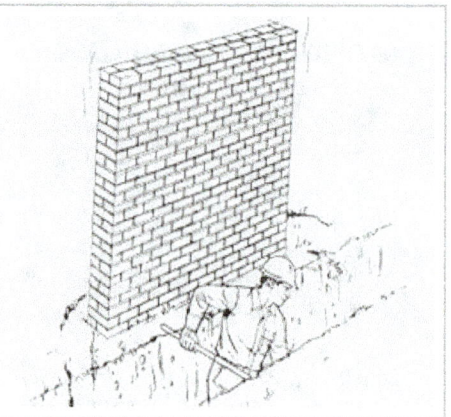
Weather Conditions:	Clear/Wet Soil	
Type of Operation:	Trenching	
Size of Work Crew:	2	
Competent Safety Monitor on Site:	No	
Safety and Health Program in Effect:	Inadeqaute	
Was the Worksite Inspected Regularly:	No, short duration	
Training and Education Provided:	Some	
Employee Job Title:	Laborer	
Age & Sex:	27-Male	
Experience at this Type of Work:	1 Year	
Time on Project:	1 Day	

BRIEF DESCRIPTION OF ACCIDENT

An employee was in the process of locating an underground water line. A trench had been dug approximately 4 feet deep along side a brick wall 7 feet high and 5 feet long. The brick wall collapsed onto the victim who was standing in the trench. The injuries were fatal.

INSPECTION RESULTS

As a result of its investigation, OSHA issued citations for violation of the standard.

ACCIDENT PREVENTION RECOMMENDATIONS

The contractor should not permit employees to excavate below the level of the base of foundation footings when walls are unpinned [29 CFR 1926.651(i)(1)]

SOURCES OF HELP

- **OSHA 2202 Construction Industry Digest** ⁻ includes all OSHA construction standards and those general industry standards that apply to construction. Order No. 029-016-00151-4, ($2.25). Available from the Superintendent of Documents, Government Printing Office, Washington DC 20402-9325, phone (202) 512-1800. Make checks payable to Superintendent of Documents. For phone orders, Visa® or MasterCard®.
- **OSHA 2254 Training Requirements in OSHA Standards and Training Guidelines** ⁻ includes all OSHA construction standards and those general industry standards that apply to construction. Order No. 029-016-00160-3, ($6.00). Available from the Superintendent of Documents, Government Printing Office, Washington DC 20402-9325, phone (202) 512-1800. Make checks payable to Superintendent of Documents. For phone orders, Visa® or MasterCard®.
- **OSHA Safety and Health Guidelines for Construction** (Available from the National Information Service, 5285 Port Royal Road, Springfield, VA 22161; (703) 605-6000 or (800) 553-6847; Order No. PB-239-312/AS, $27). Guidelines to helpconstruction employers establish a training program in the safe use of equipment, tools, and machinery on the job.

- For information on OSHA-funded free consultation services call the nearest OSHA area office listed in telephone directories under U.S. Labor Department or under the state government section where states administer their own OSHA programs.
- Courses in construction safety are offered by the OSHA Training Institute, 1555 Times Drive, Des Plaines, IL 60018, 708/297-4810.
- OSHA Safety and Health Training Guidelines for Construction (Available from the National Technical Information Service, 5285 Port Royal Road, Springfield, VA 22161; 703/487-4650; Order No. PB-239-312/AS): guidelines to help construction employers establish a training program in the safe use of equipment, tools, and machinery on the Job.

NOTE: The case here described was selected as being representative of fatalities caused by improper work practices. No special emphasis or priority is implied nor is the case necessarily a recent occurrence. The legal aspects of the incident have been resolved, and the case is now closed.

Scaffolding Accidents

Design as a Risk Factor: Australian Study, 2000–2002

- Main finding: design contributes significantly to work-related serious injury.

- 37% of workplace fatalities are due to design-related issues.

- In another 14% of fatalities, design-related issues may have played a role.

[Driscoll et al. 2008]

Photo courtesy of Thinkstock

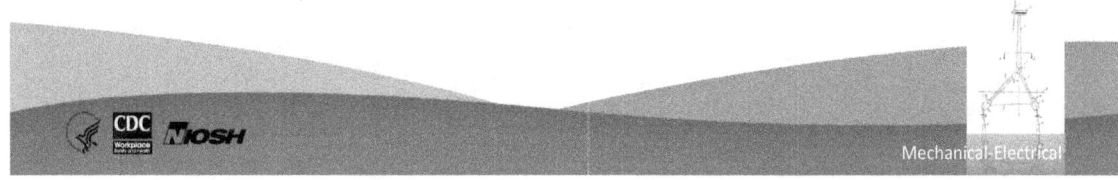

Mechanical-Electrical

NOTES

Several studies around the world have demonstrated that design can directly affect the safety of a construction site or process. The Australian government investigated the design-related root causes of their work-related fatalities. Seventy-seven (37%) of the 210 identified workplace fatalities definitely or probably had design-related issues involved. In another 29 fatalities (14%), the circumstances suggested that design issues were involved. The most common scenarios involved problems with rollover protective structures and/or associated seat belts; inadequate guarding; lack of residual current devices; inadequate fall protection; failed hydraulic lifting systems in vehicles and mobile equipment; and inadequate protection mechanisms on mobile plants and vehicles.

These fatal incidents might have been prevented if the hazards that caused them had been considered during the design phase.

SOURCES

Driscoll TR, Harrison JE, Bradley C, Newson RS [2008]. The role of design issues in work-related fatal injury in Australia. J Safety Res. 39(2):209–14 [Epub 2008:Mar 13; PubMed index for MEDLINE: 18454972].

NIOSH Fatality Assessment and Control Evaluation (FACE) Program [1983]. Fatal incident summary report: scaffold collapse involving a painter. FACE 8306 [www.cdc.gov/niosh/face/In-house/full8306.html].

Photo courtesy of Thinkstock

Fatal Incident Summary Report: Scaffold Collapse Involving a Painter

INTRODUCTION

The National Institute for Occupational Safety and Health (NIOSH), Division of Safety Research (DSR), is currently conducting the Fatal Accident Circumstances and Epidemiology (FACE) Study. By scientifically collecting data from a sample of similar fatal accidents, this study will identify and rank factors which increase the risk of fatal injury for selected employees.

On May 25, 1983, a painter suffered fatal injuries when the suspended scaffolding from which he was working collapsed. The County Coroner requested NIOSH technical assistance to develop information on factors involved with the incident data.

CONTACTS/ACTIVITIES

After receiving notification, three Division of Safety Research personnel, a safety specialist, a safety engineer, and an epidemiologist, visited at the site to interview the employer and witnesses and to obtain comparison data from suitable co-workers. The research team, the police department, and the employer examined the impounded scaffold at an independent testing laboratory.

A debriefing session was held with the employer, other employees, and the contractor. During this introductory meeting, background information was obtained about the contractor and the employer, including an overview of their safety and health program. Interviews were conducted with witnesses and co-workers. Examining the scaffold assisted the researchers in developing hypotheses about the sequence of events leading to the incident.

SYNOPSIS OF EVENTS

The two workers had placed the scaffold supporting wire rope on the 7th floor permanently installed eye hooks. They then reeved the wire rope to the scaffold stirrups which are located at each end of the scaffold staging. After reeving was complete, the workers raised the scaffolding to the 7th floor windows. This action was accomplished by turning the drive motor directional switch to the "up" position and holding the motor switch in the "on" position.

The victim had to apply caulking around the windows. After caulking half way across the floor, he had to change positions, including independent life lines with a co-worker, who survived the incident. After caulking the remaining windows, the workers switched positions again in order to begin their descent.

The co-worker stated that he turned away from the victim and faced his stirrup in preparation of descent. As he did this, he felt some movement in the scaffold. He turned and looked at the victim, who motioned by hand signal to turn the directional switch to the "down" position. The co-worker signaled "okay" and turned to face his stirrup. As he was in the process of preparing his stirrup for downward movement plus getting his lanyard grab device ready to move down, he felt several sudden jerks and was suddenly dangling from his life line. After regaining his composure, the co-worker looked for the victim in the area of his life line. The co-worker then noticed the victim lying in the street across from the building.

GENERAL CONCLUSIONS AND RECOMMENDATIONS

There is some evidence which indicates the deceased was not familiar with the operation of this type of scaffold. For this type of scaffold, the operator must operate the drill and a brake lever at the same time with one hand, while releasing his lanyard on the safety line with the other hand.

Additionally, the victim's lanyard failed to prevent the fatal fall for one of two reasons. Either the lanyard was deteriorated to the extent that the impact load was in excess of the lanyard strength or the lanyard became entangled in the scaffold components.

It is suspected that the wire rope broke because the hoist's secondary safety mechanism did not function quickly enough. The wire rope broke at a level 20+ feet below where the scaffold was originally positioned. When the mechanism finally activated, the force of the falling scaffold caused the emergency braking cam to squeeze the rope to such an extent that it actually cut 5 of the 6 strands. The remaining strand was not of sufficient strength to hold the falling scaffold and it also broke.

It is recommended that workers who use scaffolds should be trained in the proper use, maintenance, and limitations of scaffolding, life lines and lanyards. Also management should be aware of their responsibilities when their workers are using scaffolds. Safety requirements for scaffolding are outlined in the OSHAct regulations 1910.28, 1910.29 and 1926.451.

Accidents Linked to Design

- 22% of 226 injuries that occurred from 2000 to 2002 in Oregon, Washington, and California were linked partly to design [Behm 2005]

- 42% of 224 fatalities in U.S. between 1990 and 2003 were linked to design [Behm 2005]

- In Europe, a 1991 study concluded that 60% of fatal accidents resulted in part from decisions made before site work began [European Foundation for the Improvement of Living and Working Conditions 1991]

- 63% of all fatalities and injuries could be attributed to design decisions or lack of planning [NOHSC 2001]

NOTES

Research conducted in the United States, Europe, and other regions has shown that design does affect the inherent risk in constructing a facility. Research linked design to 22% of injuries that occurred in western states and 42% of fatalities across the country. European researchers found that nearly two-thirds of fatalities and injuries were linked to design. Facility designers are encouraged to consult with occupational safety and health professionals early in the design process to identify and design out hazards and to reduce risk of injury, illness, and death.

SOURCES

Behm M [2005]. Linking construction fatalities to the design for construction safety concept. Safety Sci 43:589–611.

NOHSC [2001]. CHAIR safety in design tool. New South Wales, Australia: National Occupational Health & Safety Commission.

European Foundation for the Improvement of Living and Working Conditions [1991]. From drawing board to building site (EF/88/17/FR). Dublin: European Foundation for the Improvement of Living and Working Conditions.

Falls

Falls

- Number one cause of construction fatalities
  - in 2010, 35% of 751 deaths
    www.bls.gov/news.release/cfoi.t02.htm

- Common situations include making connections, walking on beams or near openings such as floors or windows

- Fall protection is required at height of 6 feet above a surface [29 CFR 1926.760].

- Common causes: slippery surfaces, unexpected vibrations, misalignment, and unexpected loads

NOTES

Falls are the number one cause of deaths in the construction industry. In 2004, 445 (36%) of 1,234 deaths were due to falls [BLS 2006]. By contrast, of 751 deaths in the construction sector in 2010, 35% were attributed to falls [BLS 2011a]. This decline in fatalities was attributed more to the economic downturn than to any other factor [BLS 2011b].

Falls from any height can be fatal. In construction, workers are often high off the ground. For structural reasons, the taller cross-sections of W shapes are usually chosen for beams. The flanges on W shapes may be less than six inches wide. Workers walk on beams, sometimes without fall protection. Fall protection is highly recommended and often required in most scenarios involving heights. OSHA requires fall protection at a height of 15 feet above a surface during steel erection. For other construction phases, it is 6 feet [29 CFR 1926.760].

SOURCES

BLS [2006]. Injuries, illnesses, and fatalities in construction, 2004. By Meyer SW, Pegula SM. Washington, DC: U.S. Department of Labor, Bureau of Labor Statistics, Office of Safety, Health, and Working Conditions [www.bls.gov/opub/cwc/sh20060519ar01p1.htm].

BLS [2011a]. Census of Fatal Occupational Injuries. Washington, DC: U.S. Department of Labor, Bureau of Labor Statistics [www.bls.gov/news.release/cfoi.t02.htm].

BLS [2011b]. Injuries, Illnesses, and Fatalities (IIF). Washington, DC: U.S. Department of Labor, Bureau of Labor Statistics [www.bls.gov/iif/home.htm].

OSHA [2001]. Standard number 1926.760: fall protection. Washington, DC: U.S. Department of Labor, Occupational Safety and Health Administration.

 **Death from Injury**

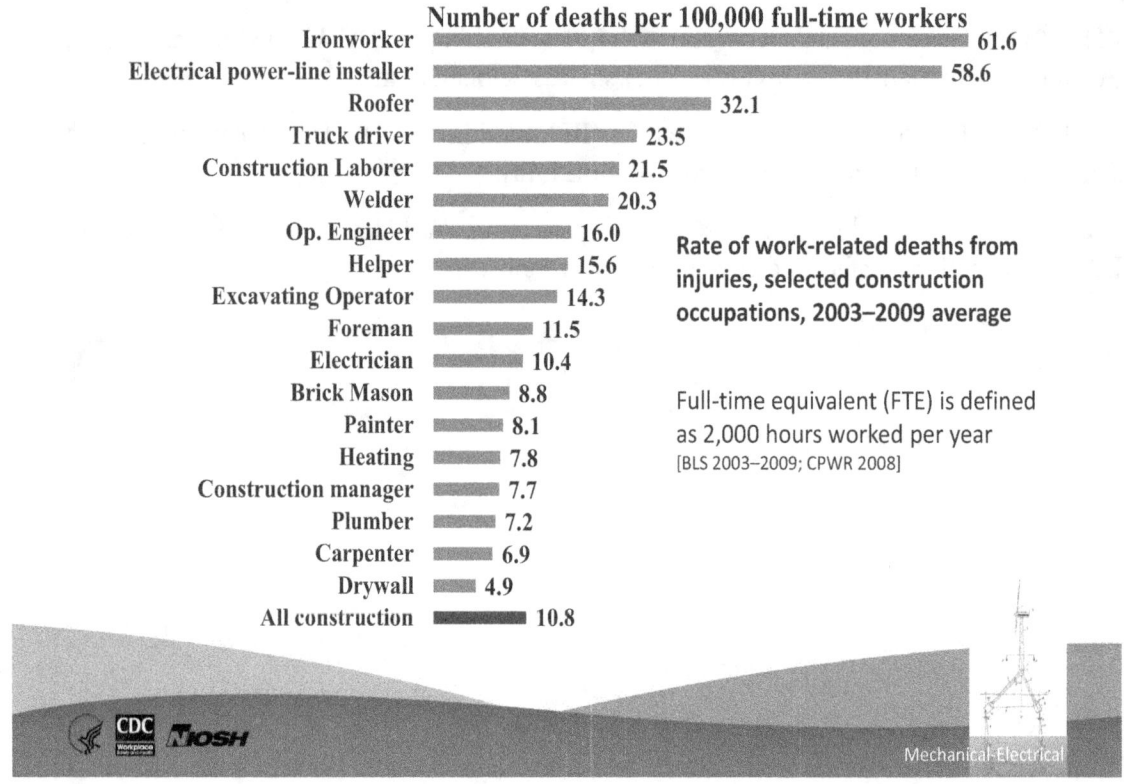

Number of deaths per 100,000 full-time workers

Occupation	
Ironworker	61.6
Electrical power-line installer	58.6
Roofer	32.1
Truck driver	23.5
Construction Laborer	21.5
Welder	20.3
Op. Engineer	16.0
Helper	15.6
Excavating Operator	14.3
Foreman	11.5
Electrician	10.4
Brick Mason	8.8
Painter	8.1
Heating	7.8
Construction manager	7.7
Plumber	7.2
Carpenter	6.9
Drywall	4.9
All construction	10.8

Rate of work-related deaths from injuries, selected construction occupations, 2003–2009 average

Full-time equivalent (FTE) is defined as 2,000 hours worked per year
[BLS 2003–2009; CPWR 2008]

Mechanical-Electrical

NOTES

The Center for Construction Research and Training compiles a "Construction Chart Book" using Bureau of Labor Statistics data [CPWR 2008]. It includes two illuminating charts useful for considering safety issues. This chart is compiled from 2003–2009 data on workplace fatalities. Ironworkers experience the highest work-related death rate, with 61.6 fatalities per 100,000 FTE.

SOURCES

BLS [2003–2009]. Census of Fatal Occupational Injuries. Washington, DC: U.S. Department of Labor, Bureau of Labor Statistics [www.bls.gov/iif/oshcfoi1.htm].

CPWR [2008]. The construction chart book. 4th ed. Silver Spring, MD: Center for Construction Research and Training.

Fatality Assessment and Control Evaluation

NIOSH FACE Program www.cdc.gov/niosh/face

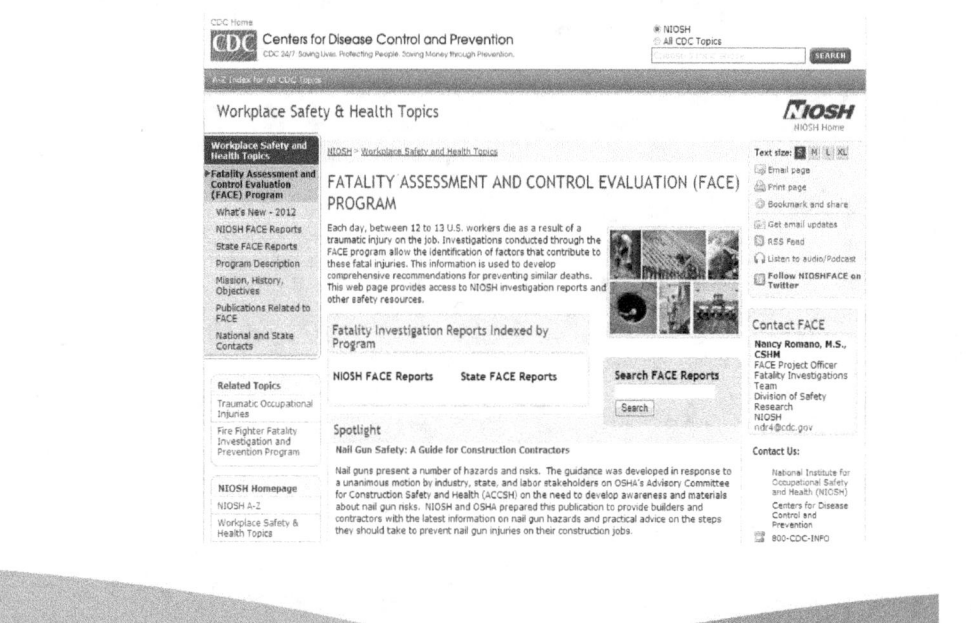

Mechanical–Electrical

NOTES

The NIOSH Fatality Assessment and Control Evaluation Program examines worker fatalities by type of injury. By studying these reports, an enterprising designer can identify recurrent problems to "design out."

SOURCE

NIOSH Fatality Assessment and Control Evaluation Program [www.cdc.gov/niosh/face/]

Death by Electrocution

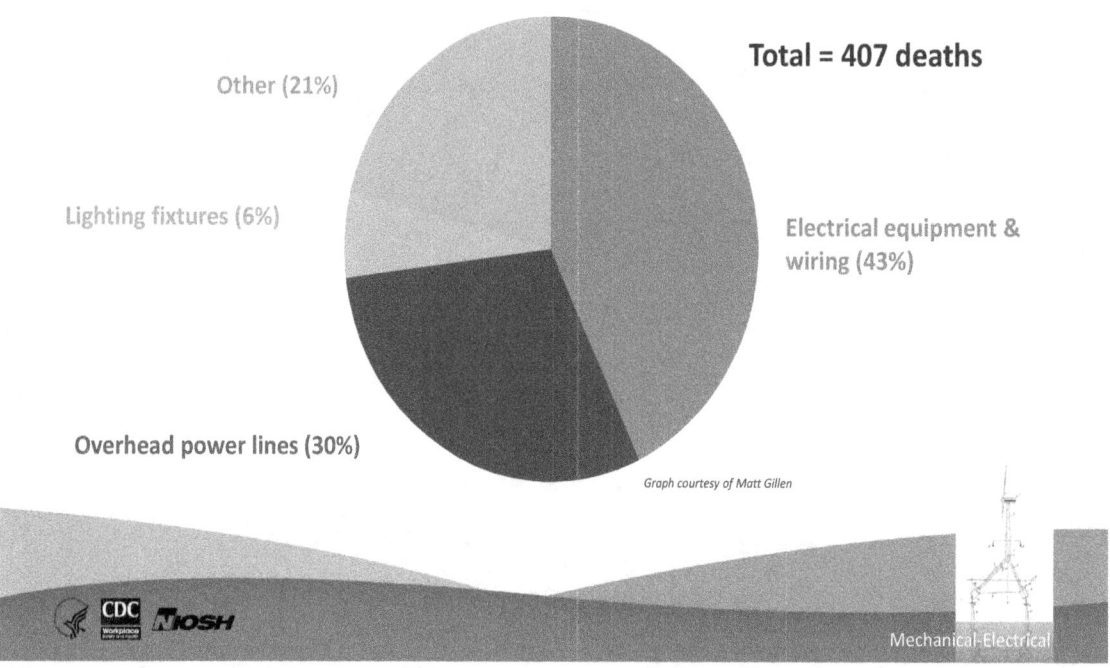

Deaths caused by contact with electricity among electrical workers in construction, total for 2003–2009 [BLS 2003-2009]

Total = 407 deaths

Other (21%)

Lighting fixtures (6%)

Electrical equipment & wiring (43%)

Overhead power lines (30%)

Graph courtesy of Matt Gillen

Mechanical-Electrical

NOTES

The top cause of death for electricians is contact with electrical equipment and wiring. Note that in addition to electrocution, death (or serious injury) can be caused by arc flashes, which can occur when a large electric current flows outside its intended path (for example, during a short circuit), passes through the air, and heats the air to temperatures as high as 35,000 degrees Fahrenheit, resulting in an explosion (arc blast).

SOURCES

BLS [2003–2009]. Census of Fatal Occupational Injuries. Washington, DC: U.S. Department of Labor, Bureau of Labor Statistics [www.bls.gov/iif/oshcfoi1.htm].

Graph courtesy of Matt Gillen

 **What is Prevention through Design?**

Eliminating or reducing work-related hazards and illnesses and minimizing risks associated with

- Construction

- Manufacturing

- Maintenance

- Use, reuse, and disposal of facilities, materials, and equipment

NOTES

PtD is a risk management technique that is being applied successfully in many industries, including manufacturing, healthcare, telecommunications, and construction. PtD is the optimal method of preventing occupational illnesses, injuries, and fatalities by designing out the hazards and risks. This approach involves the design of tools, equipment, systems, work processes, and facilities in order to reduce, or eliminate, hazards associated with work. The concept is simply that the safety and health of workers throughout the life cycle are considered while the product and/or process is being designed. The life cycle starts with concept development, and includes design, construction or manufacturing, operations, maintenance, and eventual disposal of whatever is being designed, which could be a facility, a material, or a piece of equipment.

PtD processes have been required in other countries for several years now, but in the United States PtD is being adopted on a voluntary basis. The National Institute for Occupational Safety and Health (NIOSH) is spearheading a national initiative in PtD and partnering with many professional organizations to apply the concept to their industry and professions. The Occupational Safety and Health Administration (OSHA) is very interested in PtD but is not currently considering making it mandatory.

PtD design professionals (that is, architects and/or engineers) working with the project owner (that is, the client) make deliberate design decisions that eliminate or reduce the risk of injuries or illness throughout the life of a project, beginning at the earliest stages of a project's life cycle. PtD is thus the deliberate consideration of construction and maintenance worker safety and health in the design phase of a construction project. PtD processes in construction have been required in the United Kingdom for over a decade and are being implemented in other countries such as Australia and Singapore.

PtD applies to the design of a facility, that is, to the aspects of the completed building that make a project inherently safer. PtD does not focus on how to make different methods of construction safer. For example, it does not focus on how to use fall protection systems, but it does include consideration of design decisions that influence how often fall protection will be needed. Similarly, PtD does not address how to erect safe scaffolding, but it does relate to design decisions that influence the location and type of scaffolding needed to accomplish the work. PtD concepts may also be used to design temporary structures. Some design decisions improve workplace safety. For example, when the height of parapet walls is designed to be 42", the parapet acts as a guardrail and enhances safety. When designed into the permanent structure of the building and sequenced early in construction, the parapet at this height acts to enhance safety during initial construction activities and during subsequent maintenance and construction activities, such as roof repair. In the United States, the employer is solely responsible for site safety.

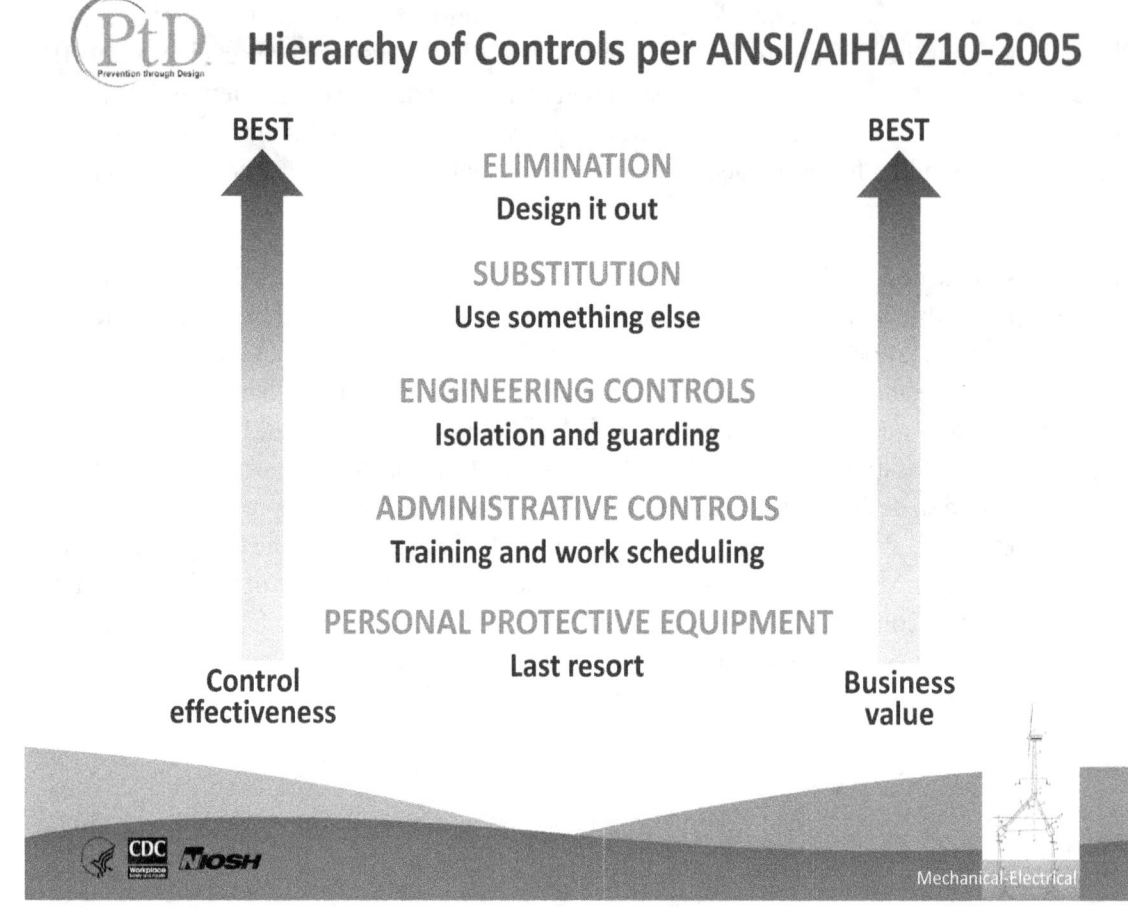

Hierarchy of Controls per ANSI/AIHA Z10-2005

BEST

BEST

ELIMINATION
Design it out

SUBSTITUTION
Use something else

ENGINEERING CONTROLS
Isolation and guarding

ADMINISTRATIVE CONTROLS
Training and work scheduling

PERSONAL PROTECTIVE EQUIPMENT
Last resort

Control
effectiveness

Business
value

Mechanical-Electrical

NOTES

This slide shows the well-accepted Hierarchy of Controls. PtD anticipates and removes potential hazardous elements at the design phase of a project through elimination or substitution. Residual risks may be minimized through the use of engineering and administrative controls.

The top of the hierarchy is better in terms of improved occupational safety and health (OSH) and cost savings. Below is a description of the different levels, from most to least effective.

Elimination: "Design out" hazards and hazardous exposures.

Substitution: Substitute less-hazardous materials, processes, operations, or equipment. A larger crane may be specified when the load or the reach approaches the crane design limit. Nontoxic chemicals are preferred. The Green Chemistry movement replaces toxic compounds with less hazardous chemicals.

Engineering controls: Isolate process or equipment or contain the hazard. Remove hazard from work zone, e.g., with exhaust ventilation. Require two hands to operate machinery. Use warning devices to warn worker about entry into hazard zone. Signs, labels, alarms, and flashing lights give warnings. Safety switches, hand guards, and other engineering controls prevent certain kinds of injuries.

Administrative controls: Job rotation, work scheduling, training, well-designed work methods, and organization are examples. Administrative controls include training modules and company procedures. A well-organized worksite is safer than a messy one. Reducing the clutter on a construction site improves worker safety by reducing the exposure to hazards. The foreman controls site layout and housekeeping policies.

Personal Protective Equipment (PPE): Includes but is not limited to safety glasses for eye protection; ear plugs for hearing protection; clothing such as safety shoes, gloves, and overalls; face shields for welders; fall harnesses; and respirators to prevent inhalation of hazardous substances.

SOURCE

ANSI/AIHA [2005]. American national standard for occupational health and safety management systems. New York: American National Standards Institute, Inc. ANSI/AIHA Z10-2005.

Personal Protective Equipment (PPE)

- Last line of defense against injury
- Examples:
  - Hard hats
  - Steel-toed boots
  - Safety glasses
  - Gloves
  - Harnesses

OSHA www.osha.gov/Publications/osha3151.html

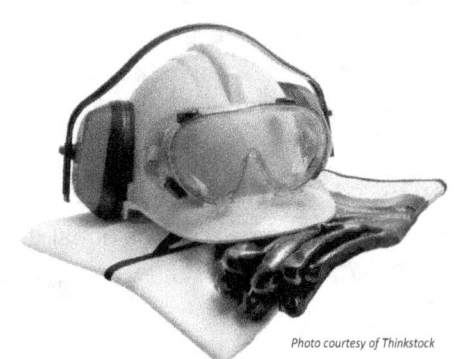

Photo courtesy of Thinkstock

Mechanical-Electrical

NOTES

Personal Protective Equipment (PPE) includes items worn as a last line of defense against injury. OSHA-required PPE can include hardhats, steel-toed boots, safety glasses or safety goggles, gloves, earmuffs, full body suits, respiratory aids, face shields, and fall harnesses.

SOURCES

CHAIR safety in design tool [2001]. New South Wales, Australia: NSW WorkCover.

OSHA PPE publications:

    www.osha.gov/Publications/osha3151.html

    www.osha.gov/OshDoc/data_General_Facts/ppe-factsheet.pdf

    www.osha.gov/OshDoc/data_Hurricane_Facts/construction_ppe.pdf

PtD Process

[Hecker et al. 2005]

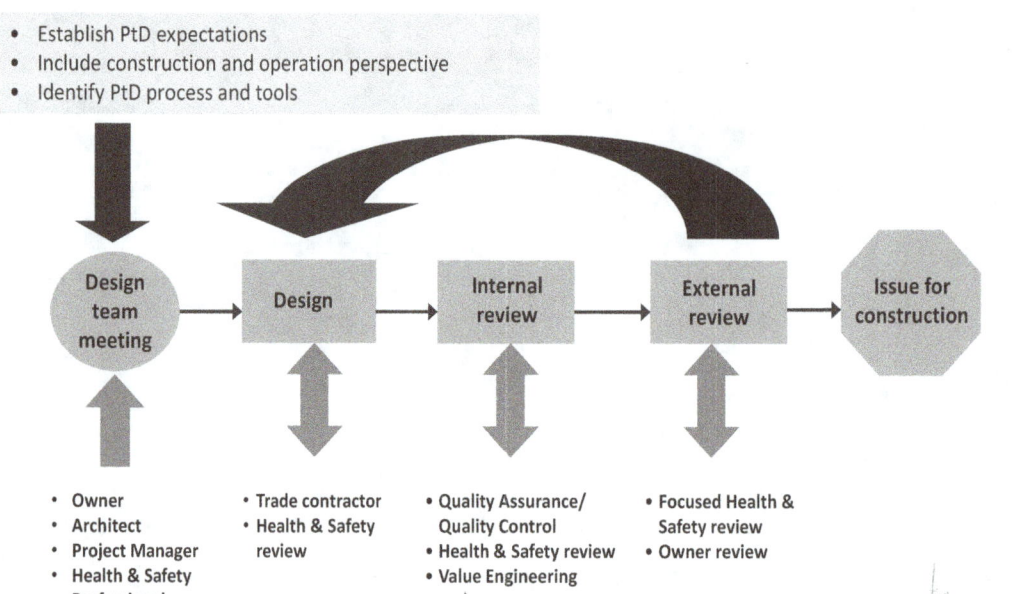

- Establish PtD expectations
- Include construction and operation perspective
- Identify PtD process and tools

| Design team meeting | Design | Internal review | External review | Issue for construction |

- Owner
- Architect
- Project Manager
- Health & Safety Professional

- Trade contractor
- Health & Safety review

- Quality Assurance/ Quality Control
- Health & Safety review
- Value Engineering review

- Focused Health & Safety review
- Owner review

NOTES

This graphic depicts the typical PtD process. The key component of this process is the incorporation of safety knowledge into design decisions. For example, site safety should be considered throughout the design process. A progress review specifically focused on site safety may be effective. Site safety knowledge can be provided by trade contractors, an on-site employee, or a hired consultant. The graphic emphasizes the importance of communication between designers and constructors. Such communication during design may reveal steps to reduce construction duration.

Many project managers schedule a Value Engineering review prior to issuing drawings for bid. The purpose is to reduce overall project costs. Unfortunately, during the review, redundant systems that are necessary to protect worker health may be eliminated. It is therefore considered a best practice to conduct a focused Health & Safety (H&S) review before drawings are issued.

SOURCE

Hecker S, Gambatese J, Weinstein M [2005]. Designing for worker safety: moving the construction safety process upstream. Prof Saf 50(9):32–44.

 **Integrating Occupational Safety and Health with the Design Process**

Stage	Activities
Conceptual design	Establish occupational safety and health goals, identify occupational hazards
Preliminary design	Eliminate hazards, if possible; substitute less hazardous agents/processes; establish risk minimization targets for remaining hazards; assess risk; and develop risk control alternatives. Write project specifications.
Detailed design	Select controls; conduct process hazard reviews
Procurement	Develop equipment specifications and include in procurements; develop "checks and tests" for factory acceptance testing and commissioning
Construction	Ensure construction site safety and contractor safety
Commissioning	Conduct "checks and tests," including factory acceptance; pre–start up safety reviews; development of standard operating procedures (SOPs); risk/exposure assessment; and management of residual risks
Start up and occupancy	Educate; manage changes; modify SOPs

NOTES

The integration of OSH goals within the design processes is an essential concept because it elevates the importance of safety and health as a value proposition in the overall design, construction, and operation of projects.

Identify hazards during conceptual design. Follow the Hierarchy of Controls to eliminate or reduce risks.

For example, how much space is needed to access, maintain, and replace HVAC units?

Use project specifications to require the inclusion of fall protection systems such as permanent anchor points for lifelines. Reduce fall hazards by specifying a ladder-free construction site.

Obtain a site plan that shows the location of existing underground and overhead utilities and develop traffic control plans to avoid those hazards.

Compare the list of desirable safety features against the detailed design.

Obtain feedback from safety and health professionals, contractors, and trade representatives. Modify the design to improve safety.

Call out required hazard controls on the drawing and in the contract specifications when possible. During procurement, compare materials and equipment received against the contract specifications. Develop a checklist for commissioning.

During construction, how do contractors communicate with the project manager and each other? Who has the authority to correct a hazardous condition on the worksite?

What procedures are followed before and after permanent equipment reaches the site? Follow the commissioning checklist!

Does the building have unusual features? Educate the owners and tenants.

Are special operating procedures required?

At each stage of the design process, think of ways to reduce the workplace risks.

Safety Payoff During Design

[Adapted from Szymberski 1997]

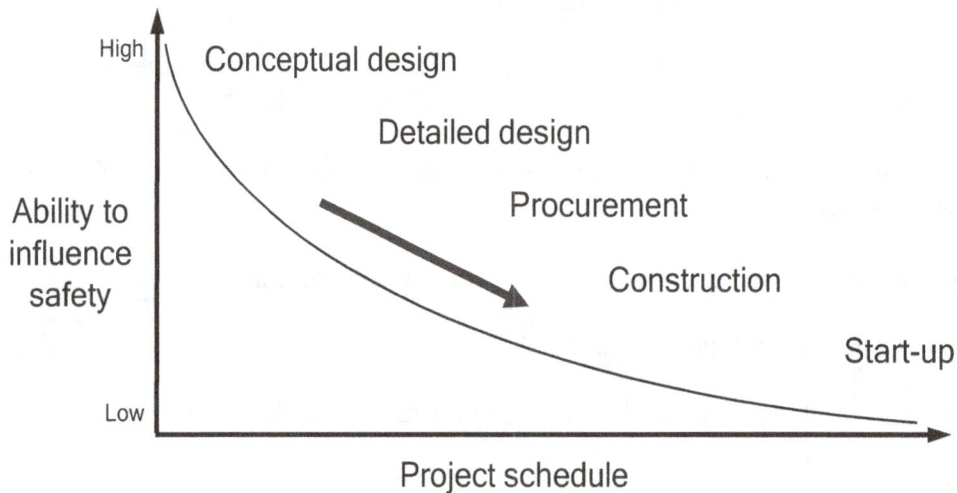

NOTES

Most owners and design professionals know intuitively that the earlier in the design process that cost is considered, the easier it is to achieve cost-effective goals. The same is true for construction duration and quality. A worker's ability to influence project criteria decreases as the design and construction progress. The same principle is true for construction safety. The earlier in the project life cycle that safety is considered, the easier it is to reduce hazards. This concept is in contrast to the prevailing methods of planning for construction site safety, which do not begin until a short time before the construction phase, when the ability to influence safety is limited.

SOURCE

Szymberski R [1997]. Construction project planning. TAPPI J *80*(11):69–74.

PtD Process Tasks

[Adapted from Toole 2005; Hinze and Wiegand 1992]

- Perform a hazard analysis

- Incorporate safety into the design documents

- Make a CAD model for member labeling and erection sequencing

Photo courtesy of Thinkstock

Mechanical-Electrical

NOTES

This slide provides more details about the PtD process. Before, during, or after the conceptual design of a building, a hazard analysis can be performed. The designer meets with field professionals to review constructability, looking through the entire design for any hazards and addressing those hazards. The field professional can teach an inexperienced designer how to minimize risks in the field.

The safety input received during conceptual design can be reflected in detailed design drawings and specifications. Another constructability review should occur as the detailed design nears completion.

Sometimes the drawings that result from a PtD process look the same as typical construction drawings, but they are inherently safer for construction. Other times, drawings include special details and labels to make it easier for workers to erect the design safely.

Construction documents can be supplemented with graphic models and tables that contribute to safe erection. For example, a CAD file can be used to label steel members for safe erection sequencing. New software such as building information modeling (BIM) is able to show the final layouts of buildings and can detect any spatial problems before construction starts. Clearly labeled shop drawings eliminate confusion during installation. The BIM program can recommend efficient, safer erection sequencing.

SOURCES

Hinze J, Wiegand F [1992]. Role of designers in construction worker safety. J Constr Eng Manage *118*(4):677–684.

Toole TM [2005]. Increasing engineers' role in construction safety: opportunities and barriers. Journal of Professional Issues in Engineering Education and Practice *131*(3):199–207.

Photo courtesy of Thinkstock

Designer Tools

- Checklists for construction safety [Main and Ward 1992]

- Design for construction safety toolbox [Gambatese et al. 1997]

- Construction safety tools from the UK or Australia
    - Construction Hazard Assessment Implication Review, known as CHAIR [NOHSC 2001]

NOTES

Most designers are not trained in PtD or construction site safety. It is therefore critical that they be given tools to facilitate the process. A PtD checklist alerts designers to common design elements that can lead to unnecessary hazards and identifies design options that are inherently safer. An example checklist is provided on the next slide.

The Design for Construction Safety Toolbox was developed by a Construction Industry Institute-sponsored research team that included leading PtD academics. This Toolbox was recently updated by Professor Jimmie Hinze at the University of Florida. The United Kingdom and Australia make available on the Web valuable PtD tools that reflect their experiences with PtD legislation and voluntary initiatives. For example, CHAIR (Construction Hazard Assessment Implication Review) is an Australian tool and methodology that systematically combines brainstorming and decisions to gradually rid the design of unnecessary hazards.

SOURCES

NOHSC [2001]. CHAIR safety in design tool. New South Wales, Australia: National Occupational Health & Safety Commission.

Gambatese JA, Hinze J, Haas CT [1997]. Tool to design for construction worker safety. J Arch Eng 3(1):2–41.

Main BW, Ward AC [1992]. What do engineers really know and do about safety? Implications for education, training, and practice. Mechanical Engineering 114(8):44–51.

Example Checklist

Item	Description
1.0	**Structural Framing**
1.1	Space slab and mat foundation top reinforcing steel at no more than 6 inches on center each way to provide a safe walking surface.
1.2	Design floor perimeter beams and beams above floor openings to support lanyards.
1.3	Design steel columns with holes at 21 and 42 inches above the floor level to support guardrail cables.
2.0	**Accessibility**
2.1	Provide adequate access to all valves and controls.
2.2	Orient equipment and controls so that they do not obstruct walkways and work areas.
2.3	Locate shutoff valves and switches in sight of the equipment which they control.
2.4	Provide adequate head room for access to equipment, electrical panels, and storage areas.
2.5	Design welded connections such that the weld locations can be safely accessed.

[Checklist courtesy of John Gambatese]

Mechanical-Electrical

NOTES

Like many PtD checklists, this example includes hazards associated with both construction and maintenance.

SOURCE

Checklist courtesy of John Gambatese

Why Prevention through Design?

- Ethical reasons

- Construction dangers

- Design-related safety issues

- Financial and non-financial benefits

- Practical benefits

Photo courtesy of Thinkstock

Mechanical-Electrical

NOTES

Engineers have strong ethical reasons to apply the PtD concept to their designs. There are practical benefits, too. Lost-time accidents delay the job, destroy crew morale, and cost money. The next few slides will show there are many reasons why owners and design professionals should be motivated to incorporate PtD in a project.

SOURCE

Photo courtesy of Thinkstock

 **Ethical Reasons for PtD**

- National Society of Professional Engineers' Code of Ethics and the American Society of Mechanical Engineers' Code of Ethics clearly states:

"Engineers shall hold paramount the safety, health and welfare of the public in the performance of their professional duties."

NSPE www.nspe.org/Ethics/CodeofEthics/index.html

ASME www.sections.asme.org/Colorado/ethics.html

NOTES

Some safety professionals and design professionals believe that PtD is an ethical duty. Nearly all national engineering societies include in their code of ethics a statement similar to the one shown here for the National Society of Professional Engineers:

"Engineers shall hold paramount the safety, health, and welfare of the public"

The American Society of Mechanical Engineers' interpretation of the Canons of the Code explicitly states:

"Engineers shall hold paramount the safety, health and welfare of the public in the performance of their professional duties."

SOURCES

American Society of Mechanical Engineers [ASME], [www.sections.asme.org/Colorado/ethics.html]

National Society of Professional Engineers [NSPE], [www.nspe.org/Ethics/CodeofEthics/index.html]

PtD Applies to Constructability

- How reasonable is the design?
  - Cost
  - Duration
  - Quality
  - Safety

Photo courtesy of the Cincinnati Museum Center www.cincymuseum.org

Mechanical-Electrical

NOTES

Most designers know that what may look great on paper might not be constructible. An important part of the design process is to evaluate the design's constructability, that is, to what extent the design can be constructed at a reasonable price, quickly, and with high quality. Safety is an important part of constructability. Accidents cost money, delay construction, and may result in bad publicity rather than acclaim for the owner.

Exciting buildings designed by creative architects require strong consideration of worker safety and health early in the design process. Owners realize these one-of-a-kind structures cost more to build and generally present unique challenges for the construction crew. Fewer construction firms have the expertise needed to build the structure, so fewer firms submit a bid, which reduces competition and therefore drives up price, resulting in higher bond and insurance costs. The timeline for procurement and construction is harder to estimate. The uniqueness of the design creates construction and maintenance challenges. Unusual materials, custom fabrications, non-standard specifications, and striking aesthetic features inherent in these designs require greater collaboration. The PtD process shown on the next slide helps the design team identify potential hazards in time to devise appropriate prevention strategies for construction crews and future maintenance workers. The project manager should include occupational safety and health professionals throughout the design process to design-in protections for workers.

SOURCE

Photo courtesy of the Cincinnati Museum Center

Business Value of PtD

- Anticipate worker exposures—be proactive

- Align health and safety goals with business goals

- Modify designs to reduce/eliminate workplace hazards in

Facilities	Equipment
Tools	Processes
Products	Work flows

 Improve business profitability!

AIHA www.ihvalue.org

NOTES

Companies that have implemented PtD programs experience lower than average injury and illness rates and lower workers' compensation expenses. However, the business value of PtD does not end there. In a study entitled Demonstrating the Business Value of Industrial Hygiene (known as The Value Study), findings showed that significant business cost savings accrue when hazards are eliminated or reduced.

SOURCE

American Institute of Industrial Hygienists [AIHA] [2008]. Strategy to demonstrate the value of industrial hygiene [www.aiha.org/votp_NEW/pdf/votp_exec_summary.pdf].

Benefits of PtD

- Reduced site hazards and thus fewer injuries

- Reduced workers' compensation insurance costs

- Increased productivity

- Fewer delays due to accidents

- Increased designer-constructor collaboration

- Reduced absenteeism

- Improved morale

- Reduced employee turnover

NOTES

PtD yields better value for owners and better health for the workers. When a project is designed with construction worker safety in mind, there are fewer hazards on site, with fewer injuries and fatalities. A reduction in injuries results in reduced workers' compensation insurance and less down-time, a direct savings for the employer. Experience shows PtD increases productivity and reduces labor costs. Safer designs lead to fewer project delays.

Industries Use PtD Successfully

- Construction companies
- Computer and communications corporations
- Design-build contractors
- Electrical power providers
- Engineering consulting firms
- Oil and gas industries
- Water utilities

  And many others

NOTES

Major corporations in diverse industries and public utilities in several states have applied PtD through initiatives or established programs. At these companies, worker safety and health are an integral part of the corporate culture. International construction firms first encountered PtD on their European projects. They brought the concepts and related cost savings home to their American operations. Many firms provide PtD training for their design engineers in the areas of construction site safety, PtD checklists, and safety constructability reviews. These firms want to hire engineers who have a basic understanding of PtD.

Electrical Hazards

MECHANICAL–ELECTRICAL SYSTEMS

Electrical Hazards

Mechanical-Electrical

NOTES

Electrocution is a concern for all workers on the site.

Working Live

"In more than half of electrical worker electrocutions, the hazard resulted because of a failure to de-energize and lock out or tag out electrical circuits and equipment. The high percentage of electrocutions caused by work on live light fixtures, especially 277 volt circuits, is especially noteworthy." [CPWR 2008]

www.elcosh.org

Mechanical-Electrical

NOTES

More than half of electrical worker deaths involve working on "live" components or wires, including live light fixtures. Remember this when you install your new chandelier.

SOURCES

United Kingdom Health and Safety Executive [UK HSE]. [2003] Electricity at work: safe working practices. p. 6. ISBN 978 0 7176 2164 4 [www.hse.gov.uk/pubns/books/hsg85.htm].

CPWR [2008]. The construction chart book. 4th ed. Silver Spring, MD: Center for Construction Research and Training [www.elcosh.org/document/1059/d000038/The%2BConstruction%2BChart%2BBook%2B4th%2BEdition.html].

OSHA Electrical Standards

29 CFR 1910.333(a)(1)

"Deenergized parts. Live parts to which an employee may be exposed shall be deenergized before the employee works on or near them, unless the employer can demonstrate that deenergizing introduces additional or increased hazards **or is infeasible due to equipment design** or operational limitations."

High School Maintenance Worker Electrocuted After Contacting a 277 Volt Electrical Cable. New Jersey FACE Investigation 95NJ070
www.cdc.gov/niosh/face/stateface/nj/95nj070.html

Mechanical-Electrical

NOTES

When you design equipment, consider ways to protect the installer and maintenance crews from the hazard of electrocution. Note that the OSHA standard includes a direct link between the need to work "live" and feasibility related to design and operational limitations.

To prevent contact with live wires, the design can provide greater capability for de-energizing select areas and for improving work practices such as pre-job planning, lockout and tagout, tool techniques, engineering controls, or use of PPE. Better electrical safety installation design has the potential to reduce the need to work "live." Read the case study and discuss design options that might have prevented this death.

SOURCES

NIOSH Fatality Assessment and Control Evaluation (FACE) Program [1995]. High School Maintenance Worker Electrocuted After Contacting a 277 Volt Electrical Cable. New Jersey FACE Investigation 95NJ070 [www.cdc.gov/niosh/face/stateface/nj/95nj070.html].

Elecrical Standard 29 CFR 1910.333(a)(1)

High School Maintenance Worker Electrocuted After Contacting a 277 Volt Electrical Cable

November 20, 1995

SUMMARY

On July 7, 1995, a 28-year-old male maintenance worker was electrocuted while working in a public school building. The incident occurred in the office area of a high school during the alteration of a wall for the construction of an alcove for a copying machine. The victim had just started the project and had removed a section of sheetrock when he discovered an electrical cable behind the wall leading from a light switch to the overhead fluorescent lights. He notified his supervisor who looked over the problem and instructed him to disconnect the power at the breaker box and to wait until he got back before he proceeded. After the supervisor left, the victim dismantled the light switch box and pulled the cable out of the wall. The victim was apparently stripping the wires on the cable when he contacted 277 volts, electrocuting him. NJDOH FACE investigators concluded that, in order to prevent similar incidents in the future, these safety guidelines should be followed:

- **Employers should develop, implement, and enforce an electrical lock-out, tag-out procedure.**
- **Employers and employees should ensure that all electrical circuits are de-energized and tested before working on them.**
- **Employers should be aware of educational and training resources for health and safety information.**

INTRODUCTION

On July 10, 1995, NJDOH FACE personnel were informed by a newspaper article of a work-related electrocution at a public high school. On July 18, 1995, FACE investigators conducted a site visit to interview the employers' representative and victim's supervisor. After viewing and photographing the scene, FACE investigators also briefly met with the investigating police detective to view the police photos and electrical cable that had been preserved as evidence. Additional information on the incident was obtained from the NJ Department of Labor Public Employees OSHA, the school's internal investigation report, and the police and medical examiner's reports.

The employer was a public high school under the jurisdiction of the regional board of education. The board of education had been in existence since 1955, with the high school opening in 1960, and employed 160 unionized workers at the time of the incident. Except for on-the-job training, the board of education did not have a specific job or safety training program for the maintenance department.

The victim was a 28-year-old maintenance worker who had been working for the school for about three weeks. He had previous experience in maintaining cryogenic liquid systems and had been a helicopter mechanic in the army. Although he had experience in servicing alarm systems, his resume did not indicate any formal electrical training.

INVESTIGATION

The incident occurred indoors at a large suburban public high school. The school had recently completed its graduation ceremonies and was out for the summer, allowing time for maintenance projects. One project was to build an alcove for the photocopy machine in one of the administrative offices. This required removing a section of sheetrock and the supporting wood studs from an office wall to construct the alcove. The victim, who had only worked for the school for a few weeks, had previously been involved with doing minor maintenance repairs under the supervision of the building and grounds foreman. Except for replacing some outlet strips, he had not been involved in doing any electrical work.

On the day of the incident, a Friday, the victim arrived for work at his usual time of 7:00 a.m. He went to work on a "hot list" of small chores, such as repairing pencil sharpeners. At about 9:00 a.m., he met with his supervisor to discuss building the photocopier alcove. He was instructed to neatly cut away the sheet rock from one side of the wall in the 12 by 12 foot office, and was left alone to do his work. The victim cut away a five by six foot area of sheetrock, exposing a BX metal-shielded electrical cable that ran horizontally through the wooden studs. The cable led from a light switch into the wall and was part of a 277 volt system for the overhead florescent lights. The victim informed his supervisor, and the two traced the cable to a junction box above the ceiling tiles. At this time, a third person (a former school maintenance person who now worked for another school) entered and the three men discussed the problem. They concluded that the cable would need to be removed and rewired away from the alcove. The plan was to pull the cable from the wall, cap the wires, and have a contractor do the rewiring. The supervisor told the victim to turn the power off in the closet (where the breaker box was), and to wait until he got back before doing anymore. He then left the room to help the former employee get some tiles.

Once again alone in the room, the victim apparently decided to go ahead with the project on his own. He first pulled the cable out of the wall switch box, shutting off the lights. He continued work by using a penlight in his mouth. A nearby secretary noticed a flash and popping sound and asked if the victim was OK, to which he smiled and said "Yes." Concerned, the secretary started to go to the school business administrator to inform him of what was going on. The victim went back to work, holding the shielded cable in his hand and using a wire stripper to remove the insulation from the wire ends. He successfully stripped one wire and was cutting through the second when he contacted the energized 277 volt conductor. At this time, another school worker heard a second popping sound and saw the victim holding the wire to his chest as he collapsed to the floor. She shouted for help and was assisted by several other workers, one of whom kicked the live wire clear of the victim. They started cardio-pulmonary resuscitation until the police, first aid squad, and paramedics arrived. The victim was transported to the local hospital where he was pronounced dead at 12:37 p.m.

The police report stated that the first pop and flash occurred when the cable contacted the metal box as it was pulled from the light switch, charring the knockout hole. The victim then contacted the electric power through the pliers, which were melted onto the wire, and was grounded through the shielded cable. The police speculated that the victim may have been stripping the wires to connect them together in order to turn the room lights back on.

CAUSE OF DEATH

The county medical examiner attributed the cause of death to electrocution. Burns were noted on the victim's hands and chest.

RECOMMENDATIONS AND DISCUSSION

Recommendation #1: Employers should develop, implement, and enforce an electrical lock-out, tag-out procedure.

Discussion: In this situation, the employer did not have a lock-out, tag-out program. It is recommended that the employer implement an effective electrical lock-out, tag-out procedure that includes de-energizing and locking out all circuits at the breaker box. All employees should receive lock-out, tag-out training and one employee should be responsible for locking out and testing the circuits. The locking out and tagging of electrical controls is required by the OSHA standard 29 CFR 1910.333 and the NJ Public Employees OSHA standard N.J.A.C. 12:100-11.

Recommendation #2: Employers and employees should ensure that all electrical circuits are de-energized and tested before working on them.

Discussion: It is not known why the victim chose to work on the energized wires after being instructed by his supervisor not to do so. To prevent future incidents, it is imperative that employers and employees de-energize all circuits that they may potentially contact. All circuits should be tested to verify that they are de-energized. It may be useful to do this with a voltage detector (such as a tic-tracer) which senses a circuit's electric field without making direct contact with the wires.

Recommendation #3: Employers should be aware of educational and training resources for health and safety information.M

Discussion: It is important that employers obtain correct information about OSHA regulations and methods of ensuring safe working conditions. Because it is often difficult for a small business to obtain this type of information, the following sources may be helpful:

NJ Department of Labor, Public Employees OSHA & U.S. Department of Labor, OSHA:

On request, NJ-PEOSHA and Federal OSHA will provide information on safety standards and requirements. PEOSHA can be contacted at the NJDOL Division of Workplace Standards, CN 386, Trenton NJ 08625, telephone (609) 292-7036. Federal OSHA has several offices in New Jersey which cover the following areas:

Hunterdon, Union, Middlesex, Warren and Somerset Counties....(908) 750-3270
Essex, Sussex, Hudson and Morris Counties...............................(201) 263-1003
Bergen and Passaic Counties...(201) 288-1700
Atlantic, Gloucester, Burlington, Mercer, Camden, Monmouth, Cape May, Ocean, Cumberland and Salem Counties....................(609) 757-5181

NJDOL OSHA Consultative Services: The New Jersey Department of Labor OSHA Consultative Service will provide free consultation to business owners on improving health and safety in the workplace and complying to OSHA standards. Their telephone number is (609) 292-3922.

New Jersey State Safety Council: The NJ Safety Council provides a variety of courses on work-related safety. There is a charge for the seminars. Their address and telephone number is 6 Commerce Drive, Cranford, New Jersey 07016, telephone (908) 272-7712

Other Sources: Trade organizations and labor unions are a good source of information on suppliers of safety equipment and training.

REFERENCES

Code of Federal Regulations 29 CFR 1926, 1991 edition. U.S. Government Printing Office, Office of the Federal Register, Washington DC.

"Control of Hazardous Energy (Lockout/Tagout)" US Department of Labor, OSHA Publication #3120, OSHA Publications Office, 200 Constitution Ave. N.W., Washington, D.C. 20210.

Overhead Power Lines

NORA Electrical Safety Goals Targeting Top Causes:

"Goal 2.1—Investigate ways to improve power line proximity warning alarms to protect operators of mobile vehicles, cranes, and nearby construction workers."

Goal 2.2—Investigate ways to protect construction workers from electrocution hazards involving power line contact through hand-carried metallic objects and vehicle-related contacts."

"Goal 2.3—Investigate ways to protect construction workers from contact with live electrical wiring and components by studying electrical installation, maintenance, and repair tasks and recommending ways to improve work practices, techniques, and tools."

NORA Construction Agenda
www.cdc.gov/niosh/nora/comment/agendas/construction/pdfs/ConstOct2008.pdf

NOTES

The National Occupational Research Agenda (NORA) has two safety goals focused on preventing fatalities caused by contact with overhead power lines. A third goal is focused on protecting workers from contact with live wires and components. Discussions on how to do this have included the use of design interventions that provide greater capabilities for de-energizing select areas. Electrical safety is an important issue to consider during safety design reviews.

SOURCES

NIOSH [2007] Preventing Worker Deaths and Injuries from Contacting Overhead Power Lines with Metal Ladders. Cincinnati, OH: U.S. Department of Health and Human Services, Centers for Disease Control and Prevention, National Institute for Occupational Safety and Health, DHHS (NIOSH) Publication No. 2007–155 [www.cdc.gov/niosh/docs/wp-solutions/2007-155/].

NIOSH [2008] NORA Construction Agenda. Cincinnati, OH: U.S. Department of Health and Human Services, Centers for Disease Control and Prevention, National Institute for Occupational Safety and Health [www.cdc.gov/niosh/nora/comment/agendas/construction/pdfs/ConstOct2008.pdf].

Site Activities

Case Study: Site Precautions to Prevent Electrocution

www.cdc.gov/niosh/face/stateface/co/94co035.html [NIOSH FACE 1994]

Mechanical-Electrical

NOTES

Read the case study and discuss the recommendations.

SOURCE

NIOSH Fatality Assessment and Control Evaluation (FACE) Program [1994]. A 35-year-old Painter Was Electrocuted When the Aluminium Ladder He Was Moving Contacted a 7,620 volt Power Line. Colorado report no. 94co035 [www.cdc.gov/niosh/face/stateface/co/94co035.html].

A 35-year-old Painter Was Electrocuted When the Aluminum Ladder He Was Moving Contacted a 7,620-volt Power Line.

SUMMARY

On July 19, 1994 several workers were spray-painting the exterior of an industrial building. The workers were using aluminum ladders to access the upper portions of the wall on which they were working. The injured worker descended his ladder, and lifted it from the wall to move it past his coworker and continue painting. As he was moving the ladder in a vertical position, it contacted a 7,620-volt power line. Another coworker hit the injured worker with both hands, knocking him from the ladder, thus breaking the electrical contact. Immediate attempts to revive the worker at the scene were unsuccessful.

The Colorado Department of Public Health and Environment (CDPHE) investigator concluded that to prevent future similar occurrences, employers should:

- **Never allow the use of aluminum ladders when the possibility of contact with overhead power lines exists.**
- **Ensure that employees request that the appropriate power company cover electrical power lines with insulating hoses or blankets if the potential for contact with lines exists.**
- **Conduct a work-site survey to assess the potential safety hazards. Once an assessment has been completed, written safety rules and procedures should be developed, implemented, and enforced.**

INVESTIGATIVE AUTHORITY

The Colorado Department of Public Health and Environment (CDPHE) performs investigations of occupational fatalities under the authority of the Colorado Revised Statutes and Board of Health Regulations. CDPHE is authorized to establish and operate a program to monitor and investigate those conditions that affect public health and are preventable. The goal of the workplace investigation is to prevent work-related injuries in the future by study of the working environment, the worker, the task the worker was performing, the tools the worker was using, and the role of management in controlling how these factors interact.

This report is generated and distributed to fulfill the Department's duty to provide relevant education to the community on methods to prevent severe occupational injuries.

INVESTIGATION

This investigation was prompted by a report to CDPHE from the Occupational Safety and Health Administration (OSHA). The investigation included interviews with the company owner and co-workers. The incident site and equipment were photographed.

The company employs one hundred people. The company has a designated safety representative and a written safety program. The safety program did not specifically address the task being performed by the deceased. The company has been in business for twelve years. The deceased had worked for the company for eight years and had been at the incident site six days.

CAUSE OF DEATH

The cause of death as determined by autopsy and listed on the death certificate was electrocution.

RECOMMENDATIONS/DISCUSSION

Recommendation #1: Aluminum ladders should never be used when the possibility of contact with overhead power lines exists.

Discussion: In this incident, the use of aluminum ladders directly contributed to the fatal injury. OSHA Regulation 29 CFR 1926.951(c)(10) prohibits the use of conductive ladders when the possibility of contact with power lines is present.

Recommendation #2: Employers should ensure that employees request that the appropriate power company cover electrical power lines with insulating hoses or blankets if the potential for contact with lines exists.

Discussion: Energized power lines in proximity to a work area constitute a significant safety hazard. Extra caution must be exercised when working in the vicinity of energized power lines. The power company should be contacted and requested to place insulating hoses or blankets on any power lines in close proximity to a work area. This protects workers who are working near power lines from making inadvertent contact.

Recommendation #3: The employer should conduct a work-site survey to assess the potential safety hazards. Once an assessment has been completed, written safety rules and procedures should be developed, implemented, and enforced.

Discussion: According to the General Duty Clause of the Occupational Safety and Health Act (Section 5 (a) 1), employers are required to provide a safe and healthy workplace for employees. To do so, employers must regularly survey the workplace to identify hazards. All identified hazards must be adequately addressed through engineering control measures or changes in work practices. Employers should also instruct each employee in the recognition and avoidance of unsafe conditions. In this and similar situations, the employer may need to provide additional training to ensure that employees understand the hazard and how to properly use equipment.

Design of Equipment

"Much can be done to improve operational safety by the careful design and selection of electrical equipment.....Circuits and equipment should be installed so that all sections of the system can be isolated as necessary.... Switch disconnectors should be suitably located and arranged so that circuits and equipment can be isolated without disconnecting other circuits that are required to continue in service." [UK HSE 2003]

NOTES

The design of an electrical system for a facility should allow maintenance functions to be performed safely. One option for large facilities is to create zones within the facility. Provide a quick disconnect switch for all circuits and equipment located in each zone of the facility.

SOURCE

United Kingdom Health and Safety Executive [UK HSE] [2003]. Electricity at work: safe working practices. p. 6. ISBN 978 0 7176 2164 4 [www.hse.gov.uk/pubns/books/hsg85.htm].

Control Panels

"...Control panels should be designed with insulated conductors and shrouded terminals so that commissioning tests, fault-finding, calibration, etc. can be carried out with a minimum of risk." [UK HSE 2003]

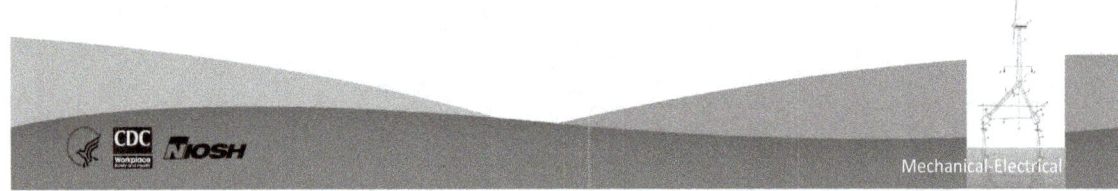

NOTES

Segregate power circuits from control circuits to reduce the risk of shock.

SOURCE

United Kingdom Health and Safety Executive [UK HSE]. [2003] Electricity at work: safe working practices. p. 6. ISBN 978 0 7176 2164 4 [www.hse.gov.uk/pubns/books/hsg85.htm].

Wind Farm

MECHANICAL–ELECTRICAL SYSTEMS

Wind Farm Case Study

NOTES

The award winning 200 MW Meadow Lake Wind Farm project in White County, IN was completed with zero lost time accidents due in part to built-in fall protection systems. The facility received two prestigious awards: The Aon Build America Award and an Indiana ACI Outstanding Achievement in Concrete Award. Students are able to relate the theory they learned in the beginning of the module to this industry example.

 **Fall Prevention**

PtD Elements for Wind Tower and Turbine
Numerous 5,000-lb. anchorage points for tie-off
Ladder fall arrest system (installed at factory)
Factory-mounted worker platforms with attached guardrails
Specially designed crane rigging attachments
Preassembly of numerous components (modular construction)
Construction sequencing to reduce workers' exposure to fall hazards
Careful planning for worker accessibility throughout the entire wind turbine structure and nacelle

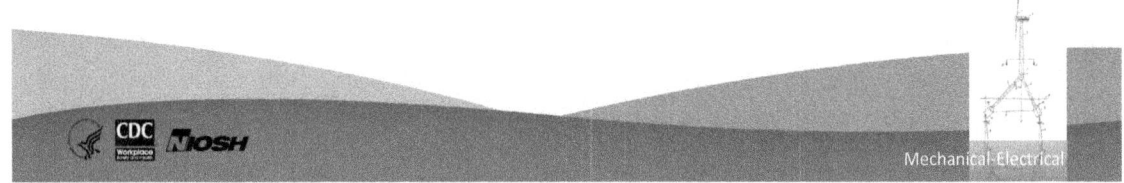

Mechanical-Electrical

NOTES

Safety and health concerns associated with windmill construction were unknown when the project was undertaken. Engineers examined safety features associated with similar structures and incorporated them into the design to prevent hazardous incidents such as falls from heights. Sections of the towers were fabricated in the shop.

Ladder Fall Arrest System

Photo courtesy of Jim McGlothlin

Mechanical-Electrical

NOTES

Inside the diameter of the wind tower, ladders were preinstalled at the factory, and fall arrest systems (note cable in the center of the ladder) were put in place. In addition, anti-fatigue rest platforms (note the platform to the left side of the ladder) were installed in each 100-foot section of the wind tower. This allows workers to step off the ladder and rest while climbing and/or performing maintenance inside the wind tower.

SOURCE

Photo courtesy of Jim McGlothlin

Crane Rigging Attachments

Photo courtesy of Jim McGlothlin

Mechanical-Electrical

NOTES

Crane rigging, with fail-safe locking mechanisms built into the lift hooks for the props, was specifically designed to lift and control (with the use of tag lines) the assembly while putting it into place on top of the wind tower. This close-up shows chokers attached to supports embedded into the propeller assembly. These supports facilitate lifting.

SOURCE

Photo courtesy of Jim McGlothlin

Propeller Accessibility Hatch

Photo courtesy of Jim McGlothlin

Mechanical-Electrical

NOTES

At the top of the tower is a large access hatch allowing workers to climb out of the generator housing to check and maintain the wind tower propellers. There are several strategically placed hooks where the worker can tie off while working at this height, to prevent falls.

SOURCE

Photo courtesy of Jim McGlothlin

 **Anchor Points**

Photo courtesy of Jim McGlothlin

Anchor points

Mechanical–Electrical

NOTES

Notice the strategically placed fall protection anchor points. The worker is tied off to a strap between his feet while working at this height to prevent falls. The most important take-home point about this PtD project is the importance of anticipating hazards in the operation and maintenance of each wind farm tower, not only during the initial construction phase but for its lifecycle. One of the reasons Bowen got the bid was because it had nested safety and health in the design of these wind towers, by pre-fabrication of the individual pieces for easy on-site assembly and in anticipation of how they would be used and maintained over time.

SOURCE

Photo courtesy of Jim McGlothlin

Nanotechnology Laboratory

MECHANICAL–ELECTRICAL SYSTEMS
Nanotechnology Laboratory

NOTES

The Birck Center for Nanotechnology at Purdue University, in West Lafayette, Indiana, is the subject of our next case study. Through the multimedia presentation, students are presented with various video clips that explain the safety features of the center. These include an uninterruptable power supply (UPS), a dual dock system for handling hazardous materials, a gas detection system, a vent system, and an acid exhaust scrubber system. In these clips, students will hear about the design of the systems. Subsequent slides identify PtD elements.

SOURCE

Captioned videos are available at www.cdc.gov.

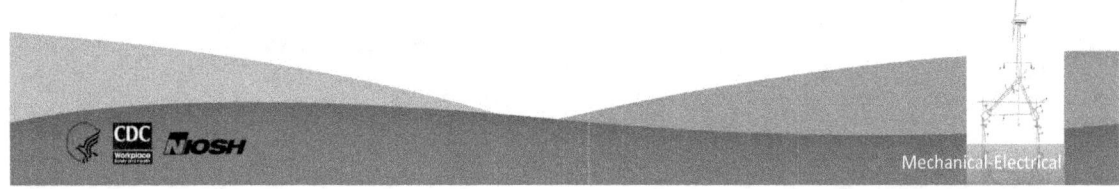

MECHANICAL–ELECTRICAL SYSTEMS
Dock Management
Nanotechnology Laboratory

NOTES

The Dock Management System protects all the occupants of the center.

 **Video of Dock Management System**

Video courtesy of Purdue University

Captioned video is available at
http://streaming.cdc.gov/vod.php?id=842fba716738e3046a3657a20ab7b5e220130730154947421

Mechanical-Electrical

NOTES

How does the dual dock system for receiving hazardous materials protect the occupants? (Watch the video!)

SOURCE

Video courtesy of Purdue University

Captioned video is available at
http://streaming.cdc.gov/vod.php?id=842fba716738e3046a3657a20ab7b5e220130730154947421

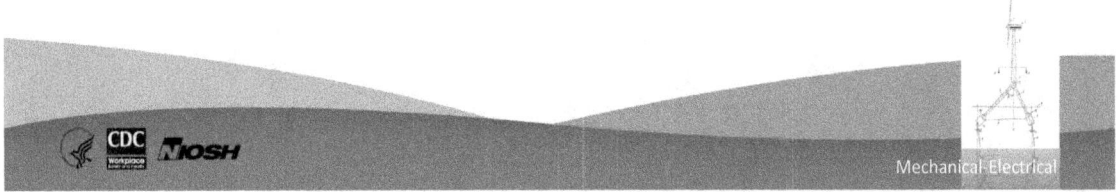

MECHANICAL–ELECTRICAL SYSTEMS
Laboratory Safety
Nanotechnology Laboratory

NOTES

Laboratories at the center were designed with the latest safety features.

Gas Storage and Monitoring System

- Ultrapure gases are distributed through stainless steel tubing.

- Hazardous gases are doubly contained, with continuous monitoring for leaks.

Photo courtesy of Purdue University

NOTES

Gases are stored in a cabinet near the loading dock used to receive hazardous materials. The distribution system is composed of stainless steel tubing. Hazardous gases are doubly contained.

SOURCE

Photo courtesy of Purdue University

 **Gas Distribution System**

- Clearly marked main gas lines run down the subfab spine

- Bulk gases stored outside the building in cabinets

- Hazardous gases stored in fireproof bunker

- All lines are supported by a chase

- Hydrogen generated on site

Photo courtesy of Purdue University

Mechanical-Electrical

NOTES

Clearly marked main gas lines run down the spine of the subfab. Secondary lines are contained in a chase, either overhead or routed below the waffle slab. Bulk gases are fed into the lines from cabinets outside the building. Hazardous gases are distributed from a fireproof bunker. Inert specialty gases may be distributed through the subfloor distribution network or from a gas cabinet located in the galley behind the laboratories. Hydrogen is generated on site as needed.

SOURCE

Photo courtesy of Purdue University

Video of Gas Detection System

Video courtesy of Purdue University

Captioned video is available at
http://streaming.cdc.gov/vod.php?id=7475cd1dc67bd951782a61474e607c4320130730155502593

Mechanical–Electrical

NOTES

Rather than storing cylinders of gases in the individual laboratory spaces, the center stores gas cylinders in cabinets in the gas room and pipes the gases into the laboratories. The Gas Detection System monitors the space in the gas storage area. Watch the video clip. Are the PtD features obvious?

SOURCE

Video courtesy of Purdue University

Captioned video is available at
http://streaming.cdc.gov/vod.php?id=7475cd1dc67bd951782a61474e607c4320130730155502593

 **Video of Chemical Spill Vent System**

Video courtesy of Purdue University

Captioned video is available at
http://streaming.cdc.gov/vod.php?id=52b2e256ceb7522018400338d040e48c20130730155232359

NOTES

Support systems include the emergency vent system that runs beneath the laboratory spaces in the subfab. (Watch the video.)

SOURCE

Video courtesy of Purdue University

Captioned video is available at
http://streaming.cdc.gov/vod.php?id=52b2e256ceb7522018400338d040e48c20130730155232359

MECHANICAL–ELECTRICAL SYSTEMS

Scrubber System
Nanotechnology Laboratory

NOTES

Now let's examine the environmental acid exhaust gas scrubber system at the center.

Exhaust Gas Scrubber System

- Provides exhaust flow for all systems where acid or base fumes and vapors may exist

- Redundant fans provide high air flow through system

- Utilizes water flowing over high-surface-area beads to remove acids and bases from air stream

- Clean air is then released into the atmosphere

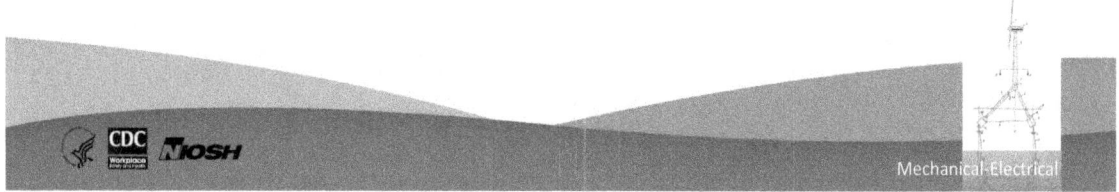

NOTES

This feature provides exhaust flow for all systems where acid or base fumes and vapors may exist. Redundant fans provide high air flow through system. Water flowing over high-surface-area beads removes acids and bases from air stream. Clean air is then released into the atmosphere.

Video of Scrubber System

Video courtesy of Purdue University

Captioned video is available at
http://streaming.cdc.gov/vod.php?id=41661e0436dde76257d454107b309dca20130730155711906

Mechanical-Electrical

NOTES

As you watch the video clip, determine what elements would be considered PtD.

SOURCE

Video courtesy of Purdue University

Captioned video is available at
http://streaming.cdc.gov/vod.php?id=41661e0436dde76257d454107b309dca20130730155711906

Evidence of PtD

- Two banks of batteries
- Monthly tests
- Specific key sequence for maintenance bypass
- Maintaining cool room temperature
- Environmental enclosure

NOTES

- There are two banks of batteries. During maintenance, one bank of batteries is always available.
- The system is designed to allow monthly testing of the backup system without causing a reset of critical systems. Conventional design would put the continuity of critical systems at risk during testing
- In order to perform a maintenance bypass in the UPS system, a specific key sequence is necessary. This prevents accidental bypass.
- The room containing the UPS system is kept at a cool temperature in order to preserve battery life.
- There is an environmental enclosure surrounding the generator in order to control the temperature, so there is no variation in startup from the heat of summer to the cold of winter.

MECHANICAL–ELECTRICAL SYSTEMS
Uninterrupted Power System
Nanotechnology Laboratory

NOTES

The uninterrupted power system is the subject of our next video. The multimedia video presentation shows the relevant safety features.

Nanotechnology Center Power System Design

- Electrical power is required to maintain safety in the facility
  - Exhaust systems
  - Makeup air systems
  - Lighting
  - Building security systems
  - Hazardous-materials monitoring systems
  - Life-safety equipment
- Utilizing PtD in the design of the power system ensures continued availability of power, even during emergency situations

[ANSI/ASSE 2011]

NOTES

In accordance with ANSI/ASSE Z590.3-2011, engineering controls at the laboratory are designed to reduce risks associated with emergency situations.

Redundant plant electrical systems are required to maintain safety in the facility. There are six systems that require electrical power. They are the exhaust and makeup air ventilation systems; lighting, security, and monitoring systems; and life safety systems. Exhaust systems are required in order to contain hazardous materials. Loss of exhaust can result in the backflow of these hazardous materials into the laboratories and cleanroom. Makeup air systems are required to function at any time when the exhaust is functioning. Loss of makeup air can affect pressure inside the building, make doors difficult to open, cause damage to the building, and present hazards to occupants. Lighting must allow safe movement within the facility. In an emergency, all exits must be clearly marked and visible to allow an orderly evacuation if warranted. Building security systems are required to keep people away from dangerous areas and to provide access for emergency responders. Hazardous-materials monitoring systems are needed to ensure indoor air quality. Many life-safety systems require electrical power. The application of PtD principles increases building safety during construction and after occupancy, even in emergency situations.

SOURCE

ANSI/ASSE [2011]. American national standard: prevention through design guidelines for addressing occupational hazards and risks in design and redesign processes. Des Plaines, IL: American Society of Safety Engineers. ANSI/ASSE Z590.3-2011.

Design of Power System

- Primary power is supplied from campus power distribution system
- Internal power distribution system within the facility provides dedicated power sources
    - Normal power
    - Sensitive power
    - Uninterrupted power
    - Emergency power

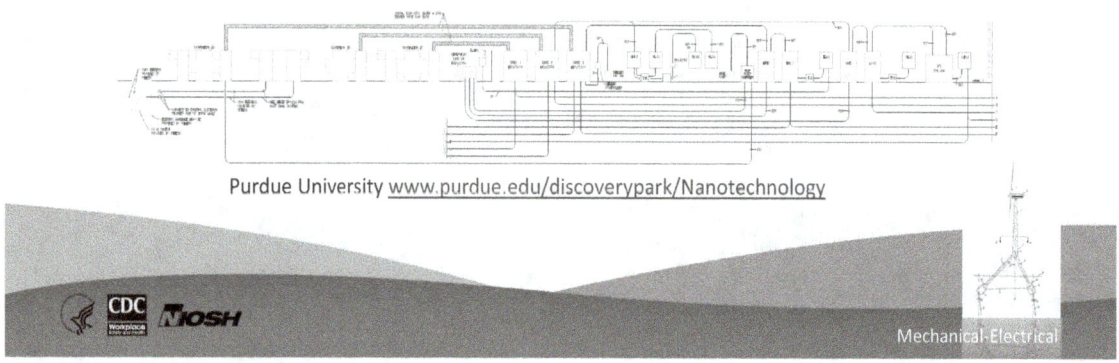

Purdue University www.purdue.edu/discoverypark/Nanotechnology

Mechanical-Electrical

NOTES

Primary power is supplied from the campus-generated power system, which is connected to the grid. This provides reliable power with a first-level backup supply. Internally, power is divided among four distribution systems. Normal power provides a pass-through of campus power, with appropriate transformation of voltages. "Sensitive power" provides isolation transformers and restrictions on equipment placed on this source to ensure minimal noise. Emergency power provides a local backup generator to ensure power when campus power is lost. Several seconds of power interruption can result as the generator starts. Uninterrupted power uses a battery source to bridge the gap between power loss and generator startup. The open transition operation allows interruptions during monthly transfer testing.

SOURCE

Diagram, videos, and slide content courtesy of Purdue University [www.purdue.edu/discoverypark/Nanotechnology]

Video of Uninterrupted Power System

Video courtesy of Purdue University

Captioned video is available at
http://streaming.cdc.gov/vod.php?id=b681f15fc233ddf4ba8eb00b5e2a004320130730155839828

NOTES

As you watch the video clip of the uninterrupted power system, try to identify the PtD elements.

SOURCE

Video courtesy of Purdue University

Captioned video is available at
http://streaming.cdc.gov/vod.php?id=b681f15fc233ddf4ba8eb00b5e2a004320130730155839828

Evidence of PtD

- Special airflow damper prevents air backflow of exhaust air during fan maintenance
- Redundant fans and pumps ensure continuous operation of system, maintaining safety inside of laboratories
- Anomalies of operation trigger text message alert to key personnel so that action can be taken prior to system failure
- System operates on emergency power
- pH and temperature are monitored to ensure proper operation of system
- "Soft" switch-over of fans to ensure continuous operation
- Critical drives kept in environmentally controlled areas

NOTES

- Regular maintenance of fans is required, but there is a potential for backflow of exhaust air from the operating fan to the fan being maintained. Installation of a special damper prevents this backflow. It is activated only during maintenance operations.

- Failure of the system would have two serious effects: loss of exhaust during operations and the potential of materials in the exhaust ductwork to backstream into the laboratories. Redundant pumps and fans prevent this eventuality.

- If an operational parameter goes into an alert status, a text message is sent to key personnel. If that status is upgraded to a failure status, a second message is sent. The redundant pump or fan has taken over, so the system is still operational but is now vulnerable. That vulnerability is therefore addressed.

- By monitoring pH and temperature, key personnel know that the system is functioning correctly.

- When a fan switch-over occurs, the operational fan ramps down while the reserve fan ramps up. This ensures that the reserve fan is operational before the operational fan shuts off.

Summary

 **Recap**

- PtD initiative is key to ensuring continuous operation of critical facility systems that guarantee the safety of those working in the facility.

- Electrical engineering elements of PtD can be understood by evaluating the building electrical distribution system.

- Mechanical engineering elements of PtD can be understood by evaluating the Exhaust Gas Scrubber system.

NOTES

PtD helps engineers design a safer workplace. The five video clips embedded in this module contain real-world examples of safety modifications in the areas of Mechanical and Electrical Engineering. Our goal is to equip the engineers of tomorrow with the knowledge to prevent injuries and save lives in the future.

 **Help make the workplace safer...**

Include *Prevention through Design* **concepts in your projects.**

For more information, please contact the National Institute for Occupational Safety and Health (NIOSH) at

Telephone: (513) 533–8302
E-mail: preventionthroughdesign@cdc.gov

Visit these NIOSH Prevention through Design Web sites:

www.cdc.gov/niosh/topics/PtD/

www.cdc.gov/niosh/programs/PtDesign/

NOTES

This presentation was intended to provide examples of construction hazards and risks that could be positively or negatively affected by design decisions. It is certainly not comprehensive in any way. All members of the construction project team (owner, designers, contractors, and safety professionals) must attempt to learn more about construction site safety early in the built environment's life cycle. The earlier more is learned, the more effective and safer the process can be. Each party has a role to play. The United Kingdom and Australia have promulgated designers' roles and responsibilities for safe construction design. Those designers are still learning how to identify and manage risks and how they can provide safer and healthier designs. We encourage the infusion of construction and safety knowledge into the design team and design reviews. Organizations and individuals seeking to positively impact construction workers' safety and health through design will need first an open mind and second a holistic view of what factors influence workers' actions and inactions. Are there any questions?

SOURCES

NIOSH PtD Web sites:

    www.cdc.gov/niosh/topics/PtD/

    www.cdc.gov/niosh/programs/PtDesign/

References

American Institute of Industrial Hygienists [AIHA] [2008]. Strategy to demonstrate the value of industrial hygiene [www.aiha.org/votp_NEW/pdf/votp_exec_summary.pdf].

ANSI/ASSE [2011]. American national standard: prevention through design guidelines for addressing occupational hazards and risks in design and redesign processes. Des Plaines, IL: American Society of Safety Engineers. ANSI/ASSE Z590.3-2011.

Behm M [2005]. Linking construction fatalities to the design for construction safety concept. Safety Sci 43:589–611.

BLS [2003–2009]. Census of fatal occupational injuries. Washington, DC: U.S. Department of Labor, Bureau of Labor Statistics [www.bls.gov/iif/oshcfoi1.htm].

BLS [2003–2009]. Current population survey. Washington, DC: U.S. Department of Labor, Bureau of Labor Statistics [www.bls.gov/cps/home.htm].

BLS [2006]. Injuries, illnesses, and fatalities in construction, 2004. By Meyer SW, Pegula SM. Washington, DC: U.S. Department of Labor, Bureau of Labor Statistics, Office of Safety, Health, and Working Conditions [www.bls.gov/opub/cwc/sh20060519ar01p1.htm].

BLS [2011a]. Census of fatal occupational injuries. Washington, DC: U.S. Department of Labor, Bureau of Labor Statistics [www.bls.gov/news.release/cfoi.t02.htm].

BLS [2011b]. Injuries, illnesses, and fatalities (IIF). Washington, DC: U.S. Department of Labor, Bureau of Labor Statistics [www.bls.gov/iif/home.htm].

CFR. Code of Federal Regulations. Washington, DC: U.S. Government Printing Office, Office of the Federal Register.

CPWR [2008]. The construction chart book. 4th ed. Silver Spring, MD: Center for Construction Research and Training.

Driscoll TR, Harrison JE, Bradley C, Newson RS [2008]. The role of design issues in work-related fatal injury in Australia. J Safety Res 39(2):209–214.

European Foundation for the Improvement of Living and Working Conditions [1991]. From drawing board to building site (EF/88/17/FR). Dublin: European Foundation for the Improvement of Living and Working Conditions.

Gambatese JA, Hinze J, Haas CT [1997]. Tool to design for construction worker safety. J Arch Eng 3(1):2–41.

Hecker S, Gambatese J, Weinstein M [2005]. Designing for worker safety: moving the construction safety process upstream. Prof Saf 50(9):32–44.

Hinze J, Wiegand F [1992]. Role of designers in construction worker safety. Journal of Construction Engineering and Management 118(4):677–684.

Lipscomb HJ, Glazner JE, Bondy J, Guarini K, Lezotte D [2006]. Injuries from slips and trips in construction. Appl Ergonomics *37*(3):267–274.

Main BW, Ward AC [1992]. What do engineers really know and do about safety? Implications for education, training, and practice. Mechanical Engineering *114*(8):44–51.

New York State Department of Health [2007]. A plumber dies after the collapse of a trench wall. Case report 07NY033 [www.cdc.gov/niosh/face/pdfs/07NY033.pdf].

NIOSH [2007]. Preventing Worker Deaths and Injuries from Contacting Overhead Power Lines with Metal Ladders. Cincinnati, OH: U.S. Department of Health and Human Services, Centers for Disease Control and Prevention, National Institute for Occupational Safety and Health, DHHS (NIOSH) Publication No. 2007–155. [www.cdc.gov/niosh/docs/wp-solutions/2007-155/].

NIOSH Fatality Assessment and Control Evaluation (FACE) Program [1983]. Fatal incident summary report: scaffold collapse involving a painter. FACE 8306 [www.cdc.gov/niosh/face/In-house/full8306.html].

NIOSH Fatality Assessment and Control Evaluation (FACE) Program [1995]. High School Maintenance Worker Electrocuted After Contacting a 277 Volt Electrical Cable. New Jersey FACE Investigation 95NJ070 [www.cdc.gov/niosh/face/stateface/nj/95nj070.html].

NIOSH Fatality Assessment and Control Evaluation (FACE) Program [1994]. A 35-year-old Painter Was Electrocuted When the Aluminium Ladder He Was Moving Contacted a 7,620-volt Power Line. Colorado report no. 94co035. [www.cdc.gov/niosh/face/stateface/co/94co035.html].

NIOSH [2008]. NORA construction agenda. Cincinnati, OH: U.S. Department of Health and Human Services, Centers for Disease Control and Prevention, National Institute for Occupational Safety and Health [www.cdc.gov/niosh/nora/comment/agendas/construction/pdfs/ConstOct2008.pdf].

NOHSC [2001]. CHAIR safety in design tool. New South Wales, Australia: National Occupational Health & Safety Commission.

OSHA [2001]. Standard number 1926.760: fall protection. Washington, DC: U.S. Department of Labor, Occupational Safety and Health Administration.

OSHA [ND]. Fatal Facts Accident Reports Index[foreman electrocuted]. Accident summary no. 17 [www.setonresourcecenter.com/MSDS_Hazcom/FatalFacts/index.htm].

OSHA [ND]. Fatal Facts Accident Reports Index [laborer struck by falling wall]. Accident summary no. 59 [www.setonresourcecenter.com/MSDS_Hazcom/FatalFacts/index.htm].

Purdue University [ND]. Birck Nanotechnology Research Center [www.purdue.edu/discoverypark/Nanotechnology].

Szymberski R [1997]. Construction project planning. TAPPI J *80*(11):69–74.

Toole TM [2005]. Increasing engineers' role in construction safety: opportunities and barriers. Journal of Professional Issues in Engineering Education and Practice *131*(3):199–207.

United Kingdom Health and Safety Executive [UK HSE] [2003]. Electricity at work: safe working practices. ISBN 978 0 7176 2164 4 [www.hse.gov.uk/pubns/books/hsg85.htm].

USC. United States Code. Washington, DC: U.S. Government Printing Office.

Other Sources

American Society of Mechanical Engineers [ASME] [www.sections.asme.org/Colorado/ethics.html]

NIOSH Fatality Assessment and Control Evaluation Program [www.cdc.gov/niosh/face/]

National Society of Professional Engineers [NSPE][www.nspe.org/ethics]

NIOSH PtD Web sites:
  www.cdc.gov/niosh/topics/PtD/
  www.cdc.gov/niosh/programs/PtDesign/

OSHA Fatal Facts Accident Reports Index [www.setonresourcecenter.com/MSDS_Hazcom/FatalFacts/index.htm]

OSHA home page [www.osha.gov/pls/oshaweb/owastand.display_standard_group?p_toc_level=1&p_part_number=1926]

OSHA PPE publications
  www.osha.gov/Publications/osha3151.html
  www.osha.gov/OshDoc/data_General_Facts/ppe-factsheet.pdf

Test Questions

1. What is the goal of PtD?

2. Give two examples of industries that have incorporated PtD into the corporate culture.

3. Name one practical benefit of PtD.

4. Give one ethical reason for PtD.

5. Give an example of hazards associated with an urban construction site.

6. List three kinds of personal protective equipment (PPE).

7. Give three reasons why PPE is considered the solution of last resort.

8. How is PtD different from engineering controls?

9. Name the players who must communicate during the design phase.

10. When in the design process is the time to consider safety?

11. Why should you visit the OSHA Web site?

12. Name three construction hazards.

13. Where can you find tools to help you create safer designs?

Answers

1. The goal of PtD is to anticipate and eliminate hazards and risks at the design phase of a project/process and to make workplaces safer for workers.

2. Construction companies, computer and communications corporations, design-build contractors, electrical power providers, engineering consulting firms, oil and gas industries, water utilities

3. Accidents on the job hurt employee morale, delay project completion, and cost money.

4. Preventable accidents should be prevented! Accidents ruin lives.

5. Examples include overhead power lines, existing infrastructure (gas, electric, and sewer), pedestrians, and traffic flow.

6. Personal Protective Equipment (PPE) includes items worn as a last line of defense against injury. OSHA-required PPE can include hardhats, steel-toed boots, safety glasses or safety goggles, gloves, earmuffs, full body suits, respiratory aids, face shields, and fall harnesses.

7. PPE is a solution of last resort because it
    a. requires the worker to wear it,
    b. may not fit because of limited size availability, and
    c. does not eliminate the hazard.

8. Engineering controls isolate the process or contain the hazard. PtD removes or reduces the hazard.

9. The entire design team must communicate, including the architect, structural engineer, civil engineer, HVAC engineer, trade representatives, and site planner.

10. Throughout!

11. OSHA regulations are updated annually. The Web site includes summaries of the latest hazard investigations. It also contains information about occupational diseases.

12. Hazards include falls, tripping hazards, falling objects, loud noises, electrocution, and musculoskeletal injuries.

13. Agencies such as OSHA and NIOSH and tools such as CHAIR can provide tools to help you create safer designs.

www.ingramcontent.com/pod-product-compliance
Lightning Source LLC
Chambersburg PA
CBHW081431170326
45166CB00008B/2388